Shifts in the Field of Mathematics Education

Editors
Peter Gates
University of Nottingham
School of Education
Nottingham
United Kingdom

Robyn Jorgensen (Zevenbergen)
Faculty of Education
Griffith University
Mt Gravatt
Queensland
Australia

ISBN 978-981-10-1162-7 ISBN 978-981-287-179-4 (eBook)
DOI 10.1007/978-981-287-179-4
Springer Singapore Heidelberg New York Dordrecht London

Softcover reprint of the hardcover 1st edition 2015

Printed on acid-free paper

Springer is part of Springer Science+Business Media (www.springer.com)

Contributors

In introducing the contributors to this book we present two voices. First, the usual descriptive account of the career and contribution each has made to the field of mathematics education. Second, and rather differently to most other books, we asked each contributor to write a very personal message about their relationship to Steve. We hope this gives you a feel for the breadth of the authorship and the personal nature of the professional relationships people have with Steve.

Jill Adler

Jill Adler is the FRF Mathematics Education Chair at the University of the Witwatersrand, and a Professor of Mathematics Education at King's College London. Her current research focuses on mathematics in and for teaching, and mathematics teacher professional development. Jill is past Vice-President of the International Commission for Mathematical Instruction (ICMI), and was on the executive of IGPME during Steve's presidency.

My friendship with Steve dates back many years and his first of many visits to South Africa as an invited speaker for the regional SAARMSTE conference in the early 1990s. Since then, we have collaborated in many mathematics education research activities. Most significant for me is the supportive role Steve played, as he worked with me to supervise my first cohort of wonderful doctoral students. It has been a privilege to write a chapter for this book, focusing in on Steve's more recent work, and how the field can and should build from this.

Marcelo Borba and Ricardo Scucuglia

Marcelo Borba is a Professor of the graduate program in mathematics education and of the mathematics department of the *Univ Estadual Paulista (UNESP)* campus of Rio Claro, where he had been the chair of the research group GPIMEM for more than 20 years. Marcelo is currently an Associate Editor of *ZDM* and a member of the editorial board of *Educational Studies in Mathematics,* where he shares duties with Steve. He has also been the editor of a collection of books in Brazil for the last 12 years. Ricardo Scucuglia completed his Ph.D. at Western University in Canada. He is currently an Assistant Professor of the Department of Education at UNESP, campus of São José do Rio Preto. Since 2000, Ricardo has collaborated with the GPIMEM group and his current research interest focusses on the notion of digital mathematical performance.

We both met Steve at PME conferences. I (Marcelo) met Steve a long while ago in the 1990s and our friendship goes back years. Steve has come to Brazil a few times and more than once he has come to work with us at UNESP. I have experienced how Steve has become an inspiration for many Brazilians who went to his university in UK as visiting scholars, post-docs, graduate students and other research collaborators.

I (Ricardo) only personally first met Steve at PME in Brazil, in 2010. But I will never forget the day Marcelo introduced me to Steve's work. I was a second year undergraduate student, attending Marcelo's class in a course on the Philosophy of Mathematics Education. That was a turning point for my views and interest in mathematics education, a fundamental aspect to subsequently becoming a member of the Research Group – GPIMEM.

Brent Davis

Brent Davis is a Professor in the Werklund School of Education at the University of Calgary, where he holds the Distinguished Research Chair in Mathematics Education. His research is focused on the educational relevance of recent developments in the cognitive and complexity sciences. The principal foci of his research are teachers' disciplinary knowledge of mathematics and the sorts of structures and experiences that might support mathematics learning among teachers.

Over the past few decades, I've encountered Steve in many places and in many ways, including through his published work, in working groups at conferences, and in shared writing and editing projects. We've even had the occasional "argument" – although that word is too strong for Steve's manner of engagement. He has an odd talent for inviting people to see the folly of their own assertions without actually pointing out inadequacies of thinking. Among his many outstanding qualities, he has more ways of saying, "That's not correct" (without saying, "That's not correct") than anyone else I know.

Paul Ernest

Paul Ernest studied mathematics, logic and philosophy at Sussex and London University. He taught mathematics at a London comprehensive school for several years and worked in teacher education at University of the West Indies, Jamaica, Homerton College, Cambridge and Bedford College. He is Emeritus Professor of at Exeter University, UK, and Visiting Professor at Brunel and Oslo Universities. His main research focus concerns fundamental questions about the nature of mathematics and how it relates to teaching, learning and society. In 'questioning the unquestionable' his views have on occasion provoked controversy. His books include *The Philosophy of Mathematics Education*, Routledge 1991, with over 1,400 citations on Google Scholar, and *Social Constructivism as a Philosophy of Mathematics,* SUNY Press, 1998. In 1990 Paul founded the *Philosophy of Mathematics Education Journal*, which continues to this day at http://www.ex.ac.uk/~PErnest/.

I met Steve in the mid-1980s when he contacted me about our shared interest in applying the philosophy of mathematics to mathematics education. For about 10 PME conferences we hung around together with a small gang of buddies laughing, learning, and not being too overawed by the great and the good. I was part of the Research into Social Perspectives of Mathematics Education group formed by Steve in the late 1980s and learned so much from our termly meetings. Steve was a great help and support when I wrote my 1991 book The Philosophy of Mathematics Education, *and even suggested the title. It's a great privilege to contribute to this book and acknowledge my debt to Steve.*

Peter Gates

Peter Gates is currently an Associate Professor in Mathematics Education at the University of Nottingham. He worked as a mathematics teacher between 1975 and 1987, including a spell in Mozambique in 1978–1979. After several years as Head of Mathematics at a very progressive school in Milton Keynes, he moved into higher education, working at The Open University (1987–1990), Bath University (1990–1993) and the University of Nottingham (1993–Present). He is an adjunct at Griffith University in Brisbane. Peter went to his first PME (10) in London in 1986, and served as a member of the International Committee and as Acting President in 2004. He is currently the Convenor of the Mathematics Education and Society group which he co-founded with Tony Cotton in 1998.

I feel like I've known Steve all my working life, and really can't remember when I first met him though it was probably at PME10 in 1986. I think I've known him long enough to remember his pony tail. Steve has always been a stimulating person to work with, talk to and argue with and I've always been really pleased that Steve was on our side. Editing this book with Robyn has been one of the highlights of my career.

Barbara Jaworski

Barbara Jaworski is Professor of Mathematics Education, Head of the Mathematics Education Centre at Loughborough University and Doctor Honoris Causa at the University of Agder, Norway. She was formerly Professor of Mathematics Education at the University of Agder and before that a Reader at the University of Oxford. She has a career that spans mathematics teaching at secondary level and first year university level, teacher education at secondary level, and teaching and supervision of doctoral students. Her research has been mainly into the development of mathematics teaching through research and through partnerships between teachers and teacher educators or didacticians. She was for 6 years Editor in Chief of the *Journal of Mathematics Teacher Education*. She is currently President of PME, the International Group for the Psychology of Mathematics Education. Her research at this time is into the development of mathematics teaching at the university level.

I have known Steve from the time we were both doing our PhDs in the late 1980s. I followed his development of theories closely and learned from what he wrote about theory. We were both part of a group called the "Understandings of Mathematics" group which met for 10 or more years; Steve was a considerate and thoughtful member of the group, always willing to debate issues in his gentle manner. I had the pleasure in 1999 of joining Steve to a conference in Taiwan which focused on Mathematics Teacher Education. We flew back together via Singapore where Steve was obliging enough to help me shop for silk fabric to make curtains. In 2000 we both contributed to an invitation conference in Norway, held in a farmhouse in a snow-covered mountainous region. The conference focused on theory in mathematics education, addressing constructivism, sociocultural theories and social practice theory. It had a rarified atmosphere in which we focused deeply on theoretical principles and differences. And we skied in the afternoons! Now, in 2014, Steve has accepted a part time professorship in the Mathematics Education Centre at Loughborough University, so we are colleagues. That pleases me very much.

Robyn Jorgensen (Zevenbergen)

Robyn Jorgensen (Zevenbergen) is currently a Professor of Education at The University of Canberra, having previously worked at Griffith University, where she worked from 1994 unitl 2014 aside from a period at Charles Sturt University (2005–2006) and then a period of time working in Central Australia as CEO and Principal of an Aboriginal Secondary Boarding College (2009–2010). Robyn has worked consistently during her academic life in the area of mathematics education and equity. Her particular foci have been with students most at risk of failing in school mathematics – low SES (or working-class) students; students from rural and/or remote regions; and Indigenous students. Within this focus, she has worked at the level of practice in order to understand the ways in which mathematics practices exclude some students and not others, and in so doing find ways to enable more students access to and success in school mathematics.

I met Steve in the early 1990s when undertaking my doctoral studies at Deakin University. He was the keynote speaker at a boutique conference organised through the University to focus on 'constructivism'. As a graduate student, it was nerve racking to sit next to Steve, a man I held in great awe, and be asked about my work. As those who have worked with Steve can attest, he has a way of talking with people that makes their work feel valued but also helps to push thinking and boundaries. He certainly helped push my thinking about constructivism and move me forward in my views about mathematics education. Over the years, we have collaborated on many projects and at conferences. I feel very honoured to have been able to work with, and collaborate with, Steve on so many projects. Working with Peter on this book has been a great pleasure and honour.

Gilah Leder

Gilah Leder is an Adjunct Professor at Monash University and Professor Emerita at La Trobe University. Her research has focussed particularly on gender issues in mathematics education, on exceptionality – predominantly high achievement – and on assessment. When Steve completed his successful term as President of the International Group for the Psychology of Mathematics Education [PME], Gilah followed in that position. Gilah is also a Past President of the Mathematics Research Group of Australasia [MERGA], a Fellow of the Academy of the Social Sciences in Australia, the recipient of the Felix Klein medal awarded for outstanding lifetime achievements in mathematics education research and development and of the 2013 MERGA Career Research Medal.

Writing a chapter for this book has been a wonderful opportunity to get to know Steve so much better and learn about the person beneath the very professional and accomplished academic.

Joao Filipe Matos

João Filipe is Professor of Education at the Institute of Education of University of Lisbon where he was President of the General Assembly until 2014. With a background in Engineering and Mathematics, he taught in primary and secondary schools in the 1970s before entering an academic career at the University. His main research interests are related to mathematical thinking, modelling and research methods in mathematics education and more recently digital technologies in education and training. João Filipe was President of PME, the International Group for the Psychology of Mathematics Education (2010–2013) and Secretary when Steve

Lerman acted as President. He was also member of the international committees of Mathematics Education and Society (MES) and Mathematical Modelling and Applications (ICTMA) international groups.

I had the pleasure and privilege of working with Steve in two research projects on mathematical thinking and learning and to appreciate his outstanding ability of analysis and co-flection. In the last 25 years we met in so many places around the world and it was always apparent that Steve has a sensitiveness and appreciation of different cultures, something which I share. His influence on my trajectory as an academic, as well as in seeking a socially and politically integrated view of the problems in education, is now very clear to me although it was at the same time smooth and powerful. I guess that with Steve I realized how the individual and the collective constitute each other.

Judith Mousley

Judy Mousley is a retired (but working) Associate Professor from Deakin University, Australia. Judy's research was mainly school based and her Ph.D. *Mathematical understanding as situated cognition* was on the nature of mathematical understanding and what teachers do to support its development. Judy was Vice President of PME, President of MERGA, Vice President of the Federation of Australian Scientific and Technological Societies, President of the Australian Mathematical Sciences Council, a member of the National Professional Standards Council – Mathematics as well as several national reviews teaching, teacher education, and mathematics curricula.

In about 1990, I attended a Discussion Group of PME on "Teachers as researchers". Steve immediately asked me to do a short presentation about action research: a typical welcoming, inclusive action for a new member. Our professional and personal friendship developed until as president of PME in 1997 he asked me to be vice president. I was stunned, but felt very honoured so accepted, and over the following few years we got to know each other well though conferences, literally hundreds of emails, and the occasional visit of Steve to Deakin University. Hence I was thrilled to be asked to write a chapter in this book.

Charalampos Sakonidis

Charalampos Sakonidis is a Professor of Mathematics Education at Democritus University of Thrace, Greece. He received his B.Sc. in Mathematics from Aristotle University of Thessaloniki, his M.Sc. from Reading University and his Ph.D. from King's College London (KCL), both in Mathematics Education. His research interests and publications are mostly related to the social and cultural aspects of teaching and learning mathematics as well as to mathematics teacher professional development.

Since the days of postgraduate studies shared at KCL and over the years of professional maturation, Steve's challenging thought, his generous ways of engaging with people, his quiet and discreet demeanour and his wonderful family have been sources of inspiration and motivation for me.

Peter Sullivan

Peter Sullivan is Professor of Science, Mathematics and Technology at Monash University. His main professional achievements are in the field of research. His recent research includes four ARC funded projects: He was a member of the Australian Research Council College of Experts for Social Behavioural and Economic Sciences for 4 years. He is an author of the popular teacher resource *Open-Ended Maths Activities: Using Good Questions to Enhance Learning* that is published in the USA as *Good Questions for Math Teaching*. Until recently he was chief editor of the *Journal of Mathematics Teacher Education*, is immediate Past President of the Australian Association of Mathematics Teachers, was the author of the Shape paper that outline the principles for the development of the Australian Curriculum in Mathematics, and was the author of the 2011 Australian Education Review on research informed strategies for teaching mathematics.

I did not have the opportunity to work with Steve until later in my research trajectory, but I was aware of his standing in the mathematics community and I read his publications much earlier. I was particularly influenced by the keynote that he delivered at the conclusion of his term as president of the International Group for the Psychology of Mathematics Education. My interest was always in classroom research. That talk connected sociocultural perspectives and the constructivist perspective that was so popular at the time with the practicalities of their implementation. Subsequently I used his thoughts and even quoted from that address in many of my applications for funding and in publications. It was therefore an honour to have been invited by Robyn Jorgensen to join a team researching the teaching of mathematics in remote Australian Indigenous communities. The specific ways that the PME address informed my subsequent research are outlined in my chapter in this book.

David Wagner

David Wagner works in the Faculty of Education at the University of New Brunswick. He is most interested in human interaction in mathematics and mathematics learning and the relationship between such interaction and social justice. This inspires his research, which has focused on positioning and authority using the lens of discourse practices, and on intercultural mathematical interactions. He currently serves on the board of directors of the journal *For the Learning of Mathematics*, the editorial board of *Educational Studies in Mathematics*, and as a member of the Nonkilling Science and Technology Research Committee. He has taught grades 7–12 mathematics in Canada and Swaziland.

My strongest image of Steven Lerman is from sitting beside him at the Agora session at the Mathematics Education and Society 6 conference in Berlin. Steve's gentle leadership was evident as he chaired the meeting with a huge stein of beer in hand. We had been encouraged to bring our drinks to the Agora, which should have been called a symposium that year because the roots of the word 'symposium' mean 'drinking together.'

Peter Winbourne

After turning his back on a career as a pilot in the RAF, Peter Winbourne taught mathematics in London comprehensive schools for 15 years. After that, he worked as an advisory teacher and for what was then the National Council for Educational Technology (NCET), and latterly, before it became defunct, the British Educational Communications Technology Agency (BECTa). He also spent a year as a tutor of beginning teachers and writer of tests at King's College. This amounts, almost exactly, to the first half of his career; he dates the second half as starting when he began working with Steve in 1993 as a senior lecturer in mathematics education.

The second half of my career (from 1993) is closely entwined with Steve's, and many details of that are to be found in, or inferred from, my chapter. I met Steve when I was working for NCET. I had heard of him before, and I had referred – critically – in my masters dissertation to his early radical constructivist writings. Like me, Steve was taken with the power of computers to provide students with tools to do mathematics, to be mathematicians in the way Papert and others had been writing about for some time. Unlike me, Steve was busy being an academic, but he found time to come to the workshops we, at NCET, organised, and to impress me with his combination of style (he wore a pony tail), humour, insight, and kindness. So, when a job came up at London South Bank, it took little to persuade me to apply (and I needed a job, anyway). I had no idea what the job actually entailed, but after 20 years, and into retirement, I think I have found out what it was about (well, enough to become a Reader in 2006). During those 20 years, my interest shifted, in no small part due to Steve's influence, from how people might use ICT in the teaching and learning of mathematics, to why anybody should ever bother to learn it at all.

Acknowledgements

This book has been a pleasure to compile. Not only do we work well together but the subject matter is something that means a great deal to both of us. So we want to start by acknowledging and thanking Steve for his friendship and contribution to our own thinking over some 30 years. We all had longer hair and more teeth when we first met!

In putting this book together though we want to acknowledge the huge contribution made by each and every author. The brief we gave them was not an easy one. We asked them to write about Steve Lerman, but also to take the discussion forward, so creating a book that had wide appeal. This book then is not merely a *Festschrift* intended to honour a respected academic, during his or her lifetime – though there was a time we felt we might be producing a *Gedenkschrift* as it was taking so long! At one point we did say to each other, "*at least let's try to get it out while we are all still alive!*" We wanted to produce a broad contribution to the field attesting to the major shifts over the past few decades. We think the contributors have achieved that – and we hope you think so too.

We want to thank Kanako Tanaka from Springer who supported us in completing the manuscript by keeping off our back as our contributors battled with their pressing deadlines and tendency to take on far too much. This book was just one more thing. Yet, because of the respect each contributor has for Steve, more pressing things were pushed aside.

We owe a huge debt of gratitude to Beryl Lerman, who surreptitiously provided us with many photographs and much inside information.

Finally, we want to thank Steve, for just being Steve.

Peter and Robyn

Contents

1 **Mapping the Field and Documenting the Contribution** 1
Peter Gates and Robyn Jorgensen (Zevenbergen)

Part I Steve the Man

2 **The Social Turn – From Up Close and Personal** 17
Peter Winbourne

3 **Steve Lerman: The Man and His Work** 29
Gilah Leder

Part II Steve Within the Field

4 **Issues of Equity and Justice in the Construction of Steve Lerman** 43
Peter Gates

5 **Tracing the Advances in the Field of Mathematics Education** 59
Charalampos Sakonidis

6 **A Speech Act in Mathematics Education – The Social Turn** 75
David Wagner

Part III Steve Photo Selection

7 **Steve Through the Years** 91
Peter Gates and Robyn Jorgensen (Zevenbergen)

Part IV Steve and International Cooperation

8 **International Research Collaborations: An Australian Perspective** 107
Robyn Jorgensen (Zevenbergen)

9 Researching the Role of the Teacher in Creating Socially Productive Classrooms that Facilitate Mathematics Learning 121
Peter Sullivan

Part V Steve's Theoretical Contributions

10 Turning Mathematical Knowledge for Teaching Social 139
Jill Adler

11 Knowledge Construction: Individual or Social? 151
Judith A. Mousley

12 Intersubjectivity in Mathematics Teaching: Meaning-Making from Constructivist and/or Sociocultural Perspectives? 171
Barbara Jaworski

13 Learning as Participatory Transformation – A Reflection Inspired by Steve Lerman's Papers and Practice 185
João Filipe Matos

14 The Philosophy of Mathematics Education: Stephen Lerman's Contributions 203
Paul Ernest

15 Lerman's Perspectives on Information and Communication Technology 215
Marcelo C. Borba and Ricardo Scucuglia

16 Troubling Mathematics "Learners" 231
Brent Davis

Index 243

Chapter 1
Mapping the Field and Documenting the Contribution

Peter Gates and Robyn Jorgensen (Zevenbergen)

Structure

In this opening chapter we want to do more than provide an overview of the book and summaries of its chapters. We want to set the scene and provide a rationale for a book celebrating the life's work and contribution of a valued academic. Through this book we want to be able, with a group of a dozen or so other colleagues, to say something about the field of mathematics education through a focus on the contribution of one key player – Stephen Lerman. We have organized the book in five parts:

Part I – Steve the Man

Chapter 2: The Social Turn – From up Close and Personal – Peter Winbourne
Chapter 3: Steve Lerman: The Man and His Work – Gilah Leder

Here two people who have known and worked with Steve for many years begin by describing his personality and his humanity, and critically how these play a most significant role in the development of his thinking. This is our first stop in a journey where we consider the role of humanity in mathematics education research. We both believe you cannot separate one's work from who one is and as a consequence we have invited some contributions specifically focusing on that. Peter and Gilah

P. Gates (✉)
School of Education, University of Nottingham, Wollaton Road, Jubilee Campus, NG8 1BB Nottingham, UK
e-mail: peter.gates@nottingham.ac.uk

R. Jorgensen (Zevenbergen)
University of Canberra, Canberra, Australia
e-mail: Robyn.Jorgensen@canberra.edu.au

P. Gates, R. Jorgensen (Zevenbergen) (eds.), *Shifts in the Field of Mathematics Education*, DOI 10.1007/978-981-287-179-4_1

both write of Steve as a (thoroughly good) human being but recognise how important it is to understand the man in order to grasp the nature of his contribution to the field.

Part II – Steve Within the Field

Chapter 4: Issues of Equity and Justice in the Construction of Steve Lerman – Peter Gates
Chapter 5: Tracing Advances in the Field of Mathematics Education – Charalampos Sakonidis
Chapter 6: A Speech Act in Mathematics Education – The Social Turn – David Wagner

Three of Steve's colleagues talk of his specific contribution to making mathematics education meaningful to the lives of young people. This is our second stop and we use it as an opportunity to think about the importance of mathematics education beyond the purely cognitive.

Part III – Steve Photograph Collection

Chapter 7: Steve Through the Years – Peter Gates and Robyn Jorgensen (Zevenbergen)

Unusually for a book of this genre, we have provided a selection of photographs of Steve. These range from a grinning 4-year-old in 1948 who is clearly up to no good, right up to his participation in a writing workshop for South African doctoral students in July 2014. We include these in order to provide very graphic evidence of the man behind the work, as described by Gilah Leder in Chap. 3:

> Steve was born in the UK, the oldest of three children. It was a close knit family, a description that applies equally well to the current generation. Thanks to some time spent as a stay-at-home father when his two daughters were quite young, Steve is particularly close to them.

Whilst we read of the social turn, the philosophy of mathematics, the contradictions within various forms of constructivism, intersubjectivity, social justice and activity theory, we should remember the boy, the brother, the father, the husband, the friend and colleague, the scholar. The trajectory Steve has followed in both his private life and his professional activity, means the fundamental values he enacts surely makes us all feel like family or friends.

Part IV – Steve and International Cooperation

Chapter 8: International Research Collaborations: An Australian Perspective – Robyn Jorgensen (Zevenbergen)
Chapter 9: Researching the Role of the Teacher on Creating Socially Productive Classrooms that Facilitate Mathematics Learning – Peter Sullivan

Steve is well known for his extensive collaborations with others, and a look through his CV illustrates the breadth of his interests. Here in our third stop two colleagues from Australia write of two specific collaborations that draw on Steve's metaphor of the zoom lens (Lerman 1998) where as researchers we are called upon to consider both the context of the classroom, but also the wider social context in which that takes place.

Part V – Steve's Theoretical Contributions

Chapter 10: Turning Mathematical Knowledge for Teaching Social – Jill Adler
Chapter 11: Knowledge Construction: Individual or Social? – Judith A. Mousley
Chapter 12: Intersubjectivity in Mathematics Teaching: Meaning-Making from Constructivist and/or Sociocultural Perspectives? – Barbara Jaworski
Chapter 13: Learning as Participatory Transformation – A Reflection Inspired by Steve Lerman's Papers and Practice – João Filipe Matos
Chapter 14: The Philosophy of Mathematics Education: Stephen Lerman's Contributions – Paul Ernest
Chapter 15: Leman's Perspective on Information and Communication Technology – Marcelo C. Borba and Ricardo Scucuglia
Chapter 16: Troubling Mathematics "Learners" – Brent Davis

For our final stop we focus on some of Steve's more theoretical contributions from several of his colleagues with whom he has worked for some time. This understandably is the longest section, given Steve's considerable theoretical contribution throughout the field. Here are accounts of his work on Bernstein's ideas, Steve's leading contribution to the constructivism era in mathematics education, a recognition that by entering the contemporary debate on technology Steve could offer perspectives on learning and policy, and finally a discussion of Steve's contribution to dealing with diversity in theoretical perspectives.

There appear to be several themes around which Steve has moved over the years and which serve to mark out decisive shifts and turns in the direction mathematics education research has moved since the early 80s. First, there is the area of teacher beliefs and the philosophy of mathematics, linking the nature of mathematics and classroom pedagogy. It was in this area Steve did his PhD in 1986 (Lerman 1986) and is discussed by Paul Ernest in Chap. 14. Second, the nature of learning and the debates between the individual and the social, opened up by an interest in constructivism in learning (see particularly Lerman 1996, 1999). These issues are addressed by Judith Mousley and Barbara Jaworski (in Chaps. 11 and 12). Third, João Filipe Matos presents a discussion on Steve's contribution to Activity Theory perspectives in mathematics education in Chap. 13. Fourth, the social role of education and particularly how the pedagogy of mathematics plays a part in the structuring role of education. Here Steve was attracted to the work of Bernstein and began a long-standing collaborated with Jill Adler who writes of this connection in Chap. 10. Fifth, the role of technology in teaching and learning in mathematics where Steve has worked with various people including Marcello Borba and Ricardo Scucuglia (who write of this collaboration in Chap. 15) something also considered by Robyn Jorgensen (Zevenbergen) in Chap. 8 (See also Crisan et al. 2006). Finally Brent Davis brings the book to a close in Chap. 16 with a discussion of Steve's assertion "*that neither complementary nor emergent views can achieve an integration*" of diverse theoretical perspectives.

In the rest of this overview chapter we take you on a journey around three main themes in Steve's work that have influenced the field of mathematics education research, and in so doing illustrate how the field develops and how one might influence that development. The three themes we explore are: how an intellectual field develops (where we draw on Bourdieu), the development of a theory of knowing (where we examine the schisms around constructivism), and finally the social turn as a social phenomenon (where we draw on Bernstein).

Locating Steve Within the Field

Any attempt to document and celebrate an individual's work needs to take into account the contextual and temporal developments that were in operation at the time of significant shifts and changes in the field. At any given time in the history of mathematics education, it is possible to see what ideas, theories and writings were of major prominence, and hence what were the major developments and themes that delineated that period from others. As key works not only mark particular epochs in mathematics education, there are certain transition points where there is a contestation of ideas and the key ideas are reshaped, or contested, or rejected in favour of new ways of thinking about knowledge and knowing. As mathematics education changes and evolves over time, different ideas or trends may have more or less value depending on what has currency at that time. One only has to consider major themes in learning theory such as behaviourism, constructivism, problem solving, or technology to see quantum shifts in what is seen as important knowledge at particular junctures in time. Being able to shift with the field, or to shape that field, is the critical work of an academic and scholar. We use these dominant ideas and the major shifts in ideas to frame this compilation of work that gives recognition and celebration to the work of Steve Lerman.

As would be expected from us, we have drawn on the work of Bourdieu to allow us to create a frame in which to locate the comprehensive work undertaken by Steve. Most notably, Bourdieu's notion of "field" is a most effective construct in which to locate the various movements in Steve's work. Coming to understand and locate Steve's work within the broader field of mathematics education, through the use of field, provides a strong framing for this compilation of work. The field, as described by Bourdieu, is

> . . .a field of forces, whose necessity is imposed on agents who are engaged in it, and as a force of struggles within which agents confront each other, with differentiated means and ends according to their position in the structure of the field of forces, thus contributing to conserving or transforming its structure. (Bourdieu, p. 32)

In positioning Steve's work within the field of mathematics education, we can see how his work at different points in time has had significant influence (and power); while at other times he has been instrumental in the transitions in the field. This is testimony to a significant scholarly enterprise, which we hope is adequately captured in the collection of chapters within this book.

Not only does one need to consider the topics of Steve's work at particular points in time, but also the credence that are given to these ideas. For within the field at any point in time, some ideas will hold more value, and hence power, than others. For those who are the purveyors of those ideas that have capital at a particular point in time, then they are likely to also have other sorts of power (such as salaries, or titles, or positions) bestowed upon them, thus making their position within the field as legitimate. This is evident when Steve's work is considered. His early work with a range of theories (such as constructivism and activity theory) and his seminal work on the 'social turn' were significant contributions to mathematics education that enabled him to gain not only formal recognition as a professor of education but also to have significant positions within mathematics education, including the prestigious presidency of the Psychology of Mathematics Education Group. This conversion of goods (or in this case intellectual resources) to other forms of capital is made possible when the framing of such intellectual enterprise is undertaken with the practices of a field.

What we want to describe here is how an analysis of Steve provides us with a very practical exemplification of what Bourdieu describes theoretically as the development of a social field. Here we have an individual playing out a brokerage role between different contested positions. As a social field develops, Bourdieu argues, differing forces vie for power. In our field, the cognitive psychologists, the developmental psychologists, the constructivists, the social constructivists, the critical theorists, have been arguing out different perspectives. Within the field, individuals stake out their ground and claim dominance. They give themselves titles to separate them out from the ordinary folk, and if being called Dr isn't enough, they can apply to be called Professor so everyone will know they are a class apart and deserve to be listened to. Whilst Steve became a professor, he did so in a second tier UK university (but one which avowedly championed the participation of the working class in higher education), rather than in the Premier League of UK "Russell Group" Universities such as Oxford, Cambridge, London, Nottingham, . . .which so blatantly privilege those with considerable social and economic capital. So how did Steve play such a dominant role in shaping and shifting the field? We think the chapters in this book provide us with some answers to this question. The fact that Steve has had so much influence on the field, despite the status of London Southbank University in relation to the Oxbridge/Redbrick/Russell Group universities in the UK says something significant about the man over the status of his institution.

A look through Steve's CV (https://sites.google.com/site/lermansteve/home/publications) illustrates the diverse paths he has taken over his career. Steve appears to have manoeuvred this path through his ability to expose and offer resolution of tensions between contested zones of our field; he does this with considerable skill, but also achieves it through his personality and his demeanor. This process is something that Bourdieu rarely articulates in his theoretical exemplifications of the development of social fields. However in this book, we maintain that we can see very clearly, how one individual plays a part in this process. Steve is rarely someone you see as domineering, forceful or self-obsessed; in fact, one rarely hears a bad word said about him! A quick look at the personal comments offered by the authors in the front of this book also attests to the profound, positive view that his colleagues hold of him. If only the same could be said about us. This is a rare accomplishment for someone who has had such power of influence within the field, to have also remained so unimposing and deeply valued to the people within that field. In our comprehensive interactions with many colleagues, we have never heard anything but positive comments about Steve – a very rare accomplishment in such a competitive and individualist domain of life.

What is clear in the stories told by the various chapter authors is the ways in which Steve has moved with/in the field, at times supporting the dominant discourses that reinforce relations of power, but at other times, challenging powerful discourses and, in so doing, transforming the field in ways that have enabled new directions of thinking. While Steve may be short in stature (a joke he often makes of himself, along with his hair-challenged head), his voluminous work is that of an intellectual giant. He has demonstrated a capacity for identifying key movements in the field and working with these to either reinforce positions or to move positions. For each of the authors in this collection, he has had a significant impact – personally and/or professionally. The collection of his oeuvre attests to his capacity to work closely with colleagues (and friends) offering a wide range of powerful ideas that have impacted on the field of mathematics education.

Steve Lerman has a position in the field of mathematics education that is recognised by the interest which many around the world hold for his writing. He has had a very colourful career with invitations for all corners of the globe to work closely with colleagues, to present his work, to lead reforms at a number of institutions, and to provide leadership and stewardship of the field through his roles in many significant organizations. His extensive CV illustrates his capacity to draw on the current moves and trends in mathematics education but also to provide insights (such as his 'social turn') that have also shaped the field. Over the 30 years of his professional life Steve has contributed in many significant ways to the field – both in terms of his conceptualization of key constructs within the field – such as theories (Bernstein, Activity theory, constructivism), research methodology, teacher education, identity, and as well as undertaking professional roles including the highly prestigious role of the peak organization with the mathematics education community – President of PME, and Chair of the British Society for Research into Learning Mathematics (BSRLM)

We have taken the occasion of Steve's "retirement" from London South Bank University as an opportunity to consider how his contribution to mathematics education can help us understand the development of our field over some 30 turbulent years and at the same time to consider what it means to be human. Not only has Steve contributed significantly to the field as a scholar, he has also shaped the personal worlds of many academics. He is a man of great intellectual integrity and a good man. The chapters in this book draw on these two significant components of Steve and his work in mathematics education. Many of the authors in this book have been personal friends of Steve while others have had their intellectual ideas shaped by his sharp mind. Whether one is a personal friend or a colleague for even (especially!) a stranger, Steve has the capacity to make people feel good about their ideas. He has been a willing partner – personal and/or intellectual – in many forays into the field of mathematics education. Each author in this book has been touched in some way by Steve and his work.

In mapping Steve's academic trajectory, it becomes quite clear that he has offered significant input into many aspects of mathematics education. More interesting is how his ideas have been shaped, many of them growing out of the other. This is most obvious when observing his early work in beliefs and attitudes, moving to constructivism, then to activity theory and then to more sociological theories. Steve has not been averse to using a range of tools to explore phenomenon in mathematics education, often being at the forefront of ideas within the field. These shifts in his theorization of mathematics education bear testimony to the ever changing nature of a field. There are clear intersections with the propositions being put forward through Steve's work at particular points in time, and those practices within the field that are either needing to reconceptualised – for example, the move from cognitive theories such as constructivism to other theories that recognized the impact of the social on the mind – which was evident in his turn to activity theory as a means to explore how meanings were being constructed by learners.

As we have shown, Steve has been a force within the field of mathematics education, not only in terms of his contribution to the field, but also his capacity to shape the field in new ways. Mathematics education as a field will have struggles in which various forms of knowledge convey status and power at a given point in time. Some players operating within the field are able to reproduce the dominant discourse positions within the field and amass what constitutes capital within it. In contradistinction, others are able to contest the status quo and create new spaces within the field drawing on their concomitant status and capital. Bourdieu (1988) describes the contestations within the field as a social space in which there is:

> both a field of forces, whose necessity is imposed on agents who are engaged in it, and as a field of struggles within which agendas confront each other, with differentiated means and ends according to their position within the structure of the field of forces, thus contributing to conserving or transforming its structure (p. 32).

It is within this conceptualization of our field that we examine Steve's contribution.

Theories of Knowing

During the 1980s and 1990s one main debate occupied a great deal of attention within the field of mathematics education research: the nature of learning as constructive or transmissive. Partly as a result of widespread interest in the work of Ernst von Glasersfeld (1983), people began to argue that learners needed to be seen as active creators of mathematics and to explore the implications for pedagogy. This developed into radical constructivism (von Glasersfeld 1984, 1995) and then social constructivism (Ernest 1998).

As discussed in Judith Mousley's chapter, we have been able to observe Steve's very challenging moves in the field where he has confronted the field of forces when he engaged with Les Steffe in the public debate in the *Journal of Research in Mathematics Education* (Lerman 1996). In this exchange, he challenged the dominance (and limitations) of radical constructivism. This was no mean feat since constructivist theory had been in ascendance for some time and was a significant force within the field. But it is not just anyone who can challenge sacred cows, particularly those with the clout that constructivism had at the time. To undertake such a serious challenge to the dominance of radical constructivism at that point in time could only be undertaken by a person who had amassed significant capital within the field in order that he be taken seriously and with credibility (see more of this in Peter Winbourne's chapter). In the period leading up to this public debate, Steve had established himself firmly as a well-informed scholar in the area of learning theory, and had already published papers on constructivism. Being so recognized enabled him to be able to speak with authority in this area. What is clear in the debate that ensued in JRME was that Steve not only had amassed sufficient capital to be in a position of dominance, but this was sufficient for him to be able to challenge the dominance of radical constructivism and embark on more social aspects of learning. He could speak with authority about constructivism but also with the strength to recognize the limitations of these theories. In so doing, he was able to move the dominance from cognitivist theories to those that were much more aligned with more social and cultural aspects of learning, thus making for a much richer theorization of learning.

Steve's (1989) early work with constructivism, both as a theory of learning and his subsequent critique of that theory, positioned him within mathematics education at a point where constructivism (and its various iterations) were held in a significant position of dominance. The shift from behaviourist and developmental theories of learning had opened up a new space for an alternative approach to learning which was filled by constructivism. Steve's writing in this area allowed him to amass capital and in so doing, created the potential for him to launch his challenge of constructivism. Bourdieu talks of being in possession of

> ...a sufficient amount of one of the different kinds of capital to be in a position to dominate the corresponding field whose struggles intensify whenever the relative value of the different kinds of capital is questioned,... that is, especially when the established equilibrium in the field of instances specifically charged with the reproduction of the fields of power is threatened. (Bourdieu 1988, p. 34)

To challenge the dominance of particular aspects of a field requires that agents within the field are able to act with authority. Not only was Steve able to offer significant challenges overtly as per the JRME papers-in which he engaged critically with the ideas of radical constructivism-his work since that time has been both overt and more covert through his work across a number of key areas. The capacity to work across a number of areas within the field required similar amassing of capital in order that he speaks with authority on a number of key movements including social interactionism, activity theory and more recently with Bernstein and some Foucauldian analysis. In the following sections, we draw on a number of these key turns in Steve's career and how he has been able to shape the field.

Steve and the Social Turn

In much of his work, Steve has explored some significant theoretical ideas – most notably the sociological work of Basil Bernstein, and the cultural psychology of Lev Vygotsky. Whilst many of us see the strength in tying our ropes to a single mast, Steve's eclecticism has been one of his strengths. He has taken Bernstein's ideas, and turned them onto mathematics education research itself as Jill Adler discusses in her chapter (see also Lerman and Tsatsaroni 2002). In doing so, he offers others the opening to understand more fully the operation of the discipline. As Michel Foucault is purported to have said (though no evidence exists in print) "My job is making windows where there were once walls". One of these windows has to be what has come to be known as the "social turn".

In both Peter Gates' and Peter Winbourne's chapters, Steve claims to be surprised at the influence of his "social turn" chapter (Lerman 2000). Looking at the extent to which it is cited and referred to, it certainly therefore had an unexpected and unintended effect. Why was this? We ask this because some of us have always turned that way! Our problem has been that whilst others turned, they did not turn in our (socially critical) direction. In his chapter, Steve claims to be largely providing an "overview of current ideas" (Lerman 2000, p. 19). He touches on Lave and Wenger, Vygotsky, Walkerdine, Bernstein and others, to help him pull together an account of various stands of thinking. This is something to which successive chapters in this book allude – Steve's capacity to clarify diverse strands of theory and to bring them together identifying the nuances of difference. Indeed, Brent Davis alludes to this skill in his chapter, specifically his capacity to identify "a nuance that separates a particular perspective from another".

David Wagner in his chapter talks of the importance of "speech acts" as declarative claims to the existence of something which actually bring into existence the very object of which they speak. This is a seductive idea and goes part way to explain the impact that the "social turn" chapter had. This represents an aspect of Steve's impact on the field. He often seems to say the right things at the right time, something we discussed above. This gives legitimacy to the idea, by bringing something into existence – as indeed he says himself (Lerman 2000, p. 23).

We might argue that while the speech claim is existential, bringing something into existence, the exact nature of the phenomenon is less clear. However, by giving a name to the phenomenon, he was able to concentrate a whole series of developments into a single phrase – and helpfully a simple, easily comprehended phrase. The problem with this as a theory is that the bringing into existence gives legitimacy for a range of formulations to come into existence. So, anyone can claim to be social now, because for all of his insights, Steve failed to account for and delineate the phenomenon; that however wasn't his intention in that article. Rather he was throwing the ball into touch for others to run with. It therefore leaves us with some questions. Why did successive authors move to a socio-cultural perspective? What problem was this setting out to resolve? How did the discourses in mathematics education change? And, most challenging of all, has this social turn made any difference to the lives of the most disadvantaged in society? Hence because these were not central questions, we can all be on the turn now because the speech act dilutes the phenomenon and allows for it to be vague and all-encompassing. Now in some ways, this is its considerable strength – and is what makes it so popular.

However, we might go further, because of its popularity in speaking to so many, Steve's speech act bought into existence the "social turn", we all *must* now be seen to be social because this becomes the legitimate stance. It is what Foucault refers to as that of which we can talk – the dominant discourse within the field; one cannot now explicitly deny the "social turn". This illustrates the power Steve can have within the field – it must be right because Steve Lerman said it. However, a look through the proceedings of any recent PME conference will highlight the apparent absence of any social turn in much work in our field; *plus ça change, plus la même chose* – something with which Steve is similarly disappointed. This indeed was the reason behind the establishment of the Mathematics Education and Society group (MES) in 1998 at whose first conference, Steve was a plenary speaker (Lerman and Tsatsaroni 1998) – 2 years before the social turn chapter appeared in print. So has the 'social turn' chapter made any real difference to the generalised discipline of mathematics education as a field of study? Maybe not, but it might well have influenced the thinking of some individuals within that field.

Once *we recognise* the social turn, regardless of the interests it serves, then it becomes much easier to resolve tension with the dominant discourse. This is perhaps an example of what Jill Adler refers to in her chapter by talking of Bernstein's "weak grammar" and we want to explore that notion a little more. Steve has had an attraction toward Basil Bernstein's ideas for some time, collaborating frequently with Jill Adler. Unfortunately Bernstein is not renowned for making his ideas communicable, quite the opposite in fact, so we are going to try to draw on his 1999 paper (Bernstein 1999) and Johan Muller's chapter on knowledge structures (Muller 2007) to examine Steve's social turn contribution to the field. Bernstein's 1999 paper, moved his earlier analysis (Bernstein 1986) much further on into an analysis of different discourse structures, important because Steve's chapter is an attempt to provide an overview drawing on several theoretical strands from within mathematics, mathematics education, social and cultural psychology. By drawing these different disciplinary discourses together Steve has produced a synthesis that builds on several different rule systems – (mathematical, sociological, psychological, cultural) and in Bernstein's terms this produces a generalised discourse with very weak internal structure – what Bernstein terms a "weak grammar" where the internal properties of the knowledge structure is largely unspecified (Muller 2007, 71). Bernstein argues this comes about through the way the knowledge, competencies and literacies are segmented (Bernstein 1999, 161) what he terms a "horizontal discourse" (See Jill Adler's chapter for more on this). This does not make Steve's argument incoherent or inconsistent, but does allow for a much wider – thus weaker – set of discourses to proliferate.

By drawing on a system with a weak set of rules for what can be said and how we say it, leaves open the possibility of contrasting interpretation and truth claims. This "weak grammar" and "horizontal discourse" are weaknesses of the field of mathematics education such that much work within it still obscures the way in which social class stratification is fostered, or at least not challenged; much obscures the social rather than examines it, but of course that rather depends on what you call "the social". To some in mathematics education, the social refers to the organisation of the classroom and our interpretations of interactions therein. For example Paul Cobb writes of the importance of a situated view in the same volume in which Steve's social turn chapter appears (Cobb 2000). By drawing on his work with Erna Yackel on socio-mathematical norms (Yackel and Cobb 1996), the focus of Paul's closely detailed analysis is very much on the classroom as the unit of analysis.

There are some (and we are two) who would not consider this the "social" at all, rather it is the "interpersonal"; but "the interpersonal turn" does not have the same ring to it. Taking this alternative line, the social requires a more political approach recognising the influence on learning mathematics of economic inequality and social class stratification. In the same volume yet again, Michael Apple argues for this rather different view of the social by acknowledging that

> there is not a long tradition of mathematical education critically and rigorously examining the connections between mathematics as an area of study and the larger relations of unequal economic and political and cultural power. (Apple 2000, p. 10)

There are others though whose work takes the social turn that much further – Eric Gutstein (2006) and Ole Skovsmose (2012) in particular take the social turn further by moving outside the classroom, even outside of the school, in order to understand and counter the exclusion and underachievement of particular groups of children. For example, in disadvantaged areas teachers of small children know a major impediment to learning comes as a result of children being undernourished. Yet rarely does the social turn consider the influence of having no breakfast on learning mathematics – even now, that we have all turned social. What we might benefit from now is an understanding of exactly what developments in the field the "social turn" itself led to and has allowed to proliferate. That perhaps might be Steve's next project. The "social turn" though is a significant event, and is one of Steve's contributions of which he is righty proud (See Gilah Leder's chapter).

Steve closes his social turn chapter with the suggestion that we "search for a suitable metaphor for mind-in-society-in-mind" (p. 38). Such a metaphor we would argue already exists, but it resides in a corpus of work Steve rarely delves into – the work of Pierre Bourdieu and his exemplification of the habitus. An alternative analysis of the social turn then might have looked at the development of the field in a *sociological* sense by identifying class interests and the operation of political power within and through mathematics education.

Within the development of the field of mathematics education, Steve's work has been crucial, it has been necessary, but not sufficient to mould and shape the social turn in which significant changes to the distribution of success and power might be bought about. It has legitimised a movement, and interpellated others into the movement. Charalampos Sakonidis refers to this process in his chapter and in so doing identifies some of the key themes within the social turn – that learning is about participating in social practices and that research often focuses on the interactions between individuals within a mathematics classroom as the social. In addition there is a concern for the social injustices and social reproduction caused or exacerbated by the school.

Although much of Steve's work resides in developing and articulating theoretical positions, one recent paper (Lerman 2011) shows Steve working on data and raising deep political questions regarding the experience of exclusion, felt by some student teachers from disadvantaged backgrounds. It is clear to those of us who know Steve, that he is committed to equity and social justice as we can see from a quote indicating Steve's humanity:

Finally, we consider that equity and inclusion are aspects of mathematics education that should be of great concern to all of us, given the role of a success in school mathematics as a gatekeeper to so many fields. We believe that the social turn and the proliferation of social theories have enabled us to examine and research equity issues in ways that our previous theoretical frameworks did not allow (Lerman 2006, 12).

This is important, because whilst much of Steve's work lies outside an equity framework, this is still what drives him on (see Peter Gates' chapter).

Finally

Early on in this chapter we asked how it came about that Steve could make such a contribution to the field. Along with other contributors to this book we would argue it is his attention to detail (even "fastidious" attention as Brent Davis claims) and his ability to see the way that tensions and contradictions can be identified and balanced. He has the ability to write and speak clearly in a way that invites his audience in and gives them the impression he actually wants to talk *to* them not *past* them – a skill so many academics lack! He is self-effacing and modest, and from our own experience with Steve, he is genuine. Now these qualities are not just the sycophancy one might expect in a book testifying to an individual's contribution. They are exactly the essential skills one needs to build capacity, to encourage others to experiment with their thinking, and to encourage a thousand flowers to grow in our field. They are also the qualities that have given Steve his authority in the field. They are personal qualities we can all learn from.

Inevitably, there will be gaps in a book such as this; some potential authors were invited and for various reasons were unable to contribute. What we do have here though, is a group of people whose professional – and in some cases personal – lives have been changed through their contact with Steve. Some voices are missing from this collection. We have not for example sought out any of the young people who were taught mathematics by Steve in Israel, London and elsewhere. However, we would expect them to say something similar to most of our contributors... that Steve was a good teacher who cared about them. Because, after all, that is what we are in the business of – improving the quality of mathematics teaching and ultimately improving the lives of young people.

We have both been privileged to be close friends of Steve, since 1986 (Peter) and 1992 (Robyn). We have tried hard though to avoid a sycophantic admiration of Steve and rather provide a book that will prove a stimulating and significant addition to the literature for some time to come – as befits Steve's contribution. Of course, whether or not we have achieved that will depend on your reaction, but we do hope you enjoy the book.

References

Apple, M. (2000). Mathematical reform through conservative modernisation? Standards, markets and inequality in education. In J. Boaler (Ed.), *Multiple perspectives on mathematics teaching and learning* (pp. 243–259). Westport: Ablex Publishing.

Bernstein, B. (1986). On pedagogic discourse. In J. Richardson (Ed.), *Handbook of theory and research for the sociology of education*. New York: Greenwood Press.

Bernstein, B. (1999). Vertical and horizontal discourse: An essay. *British Journal of Sociology of Education, 20*(2), 157–173.

Bourdieu, P. (1988). *Practical reason: A theory of action*. Stanford: Stanford University Press.

Cobb, P. (2000). The importance of a situated view of learning to the design of research and instruction. In J. Boaler (Ed.), *Multiple perspectives on mathematics teaching and learning* (pp. 45–84). Westport: Ablex Publishing.

Crisan, C., Lerman, S., & Winbourne, P. (2006). Mathematics and ICT: A framework for conceptualising secondary school mathematics teachers' classroom practices. *Technology, Pedagogy and Education, 16*(1), 21–39.

Ernest, P. (1998). *Social constructivism as a philosophy of mathematics*. Albany: SUNY Press.

Gutstein, E. (2006). *Reading and writing the world with mathematics: Towards a pedagogy for social justice*. New York: Routledge.

Lerman, S. (1986). *Alternative perspectives on the nature of mathematics and the possible influence on the teaching of mathematics*. Unpublished PhD thesis, Kings College, University of London.

Lerman, S. (1989). Constructivism, mathematics and mathematics education. *Educational Studies in Mathematics, 20*(2), 211–223.
Lerman, S. (1996). Intersubjectivity in mathematics learning: A challenge to the radical constructivist paradigm? *Journal for Research in Mathematics Education, 27*(2), 133–150.
Lerman, S. (1998). A moment in the zoom of a lens: Towards a discursive psychology of mathematics teaching and learning. In A. Olivier & K. Newstead (Eds.), *Proceedings of the 22nd conference of the International Group for the Psychology of Mathematics Education* (Vol. 1, pp. 66–81). Stellenbosch: University of Stellenbosch.
Lerman, S. (1999). A response to Steffe's reply to Lerman on intersubjectivity: A case of interpretations of 'social'. *Chreods, 13*, 17–25.
Lerman, S. (2000). The social turn in mathematics education research. In J. Boaler (Ed.), *Multiple perspectives on mathematics teaching and learning* (pp. 19–44). Westport: Ablex Publishing.
Lerman, S. (2006). Theories of mathematics education: Is plurality a problem? *ZDM, 38*(1), 8–13.
Lerman, S. (2011). *Agency and identity: Mathematics teachers' stories of overcoming disadvantage*. Paper presented to the Mathematics Education and Contemporary Theory Conference, Manchester Metropolitan University. http://www.esri.mmu.ac.uk/mect/papers_11/lerman.pdf
Lerman, S., & Tsatsaroni, A. (1998). Why children fail and what mathematics education studies can do about it. The contribution/role of sociology. In P. Gates & A. Cotton (Eds.), *Proceedings of the first mathematics education and society conference*. Nottingham. http://www.nottingham.ac.uk/csme/meas/plenaries/lerman.html
Lerman, S., & Tsatsaroni, A. (2002). Developing theories of mathematics education research: The ESM story. *Educational Studies in Mathematics, 51*(1–2), 23–40. doi:10.1023/a:L1022412318413.
Muller, J. (2007). On splitting hairs: Hierarchy, knowledge and the school curriculum. In F. Christie & J. Martin (Eds.), *Language, knowledge and pedagogy* (pp. 65–86). London: Continuum.
Skovsmose, O. (2012). Toward a critical mathematics education research programme. In O. Skovsmose & B. Greer (Eds.), *Opening the cage. Critique and politics of mathematics education* (pp. 343–368). Rotterdam: Sense Publishers.
von Glasersfeld, E. (1983). Learning as a constructive activity. In J. C. Bergeron & N. Herscovics (Eds.), *Proceedings of the 5th annual meeting of the North American Group of Psychology in Mathematics Education* (Vol. 1, pp. 41–101). Montreal: PME-NA.
von Glasersfeld, E. (1984). An introduction to radical constructivism. In P. Watzlawick (Ed.), *The invented reality* (pp. 17–40). New York: Norton.
von Glasersfeld, E. (1995). *Radical constructivism: A way of knowing and learning*. London: RoutledgeFalmer.
Yackel, E., & Cobb, P. (1996). Sociomathematical norms, argumentation and autonomy in mathematics. *Journal for Research in Mathematics Education, 27*, 458–477.

Part I
Steve the Man

Chapter 2
The Social Turn – From Up Close and Personal

Peter Winbourne

My aim in this chapter is to provide something of a map of the history of the social turn, as seen from up close in Steve's thinking and writing, and reflected in his practice.

So, there is, I hope, a double loop to be found here: how Steve's writing has been an articulation of sociocultural approaches; how it has also been a reflection of sociocultural approaches to be seen in his practice.

Methodology

Writing a chapter in a book of this kind about Steve's example, obliges me to include some methodological discussion. I would like to say that this has been a longitudinal ethnographic study with Steve, and the academic and professional communities within which he has been working, as subject. Had a claim like this been at all justified, the study would certainly have been longitudinal: I have been working with Steve as a very close colleague, and friend, for 19 years. But, how close is close? For a number of years Steve and I shared an office. That office also served as a changing room and a place for me to keep my sweaty cycle gear, and also, quite regularly, as a place for Steve to keep his running gear on the days he chose to run home. For most of that time there was a pot of poisonously strong coffee always on the brew, and, though it is now some 13 years since we have had our own offices, the smell and taste linger. For the past 3 years we have been housed in a brand new building. Steve's office and mine are two doors away. I pass his door everyday as I arrive (usually after changing out of my cycling gear in the nice new changing room on the same floor). Most often he is there before me. When he is

P. Winbourne (✉)
London South Bank University, London, UK
e-mail: winboupc@lsbu.ac.uk

P. Gates, R. Jorgensen (Zevenbergen) (eds.), *Shifts in the Field of Mathematics Education*, DOI 10.1007/978-981-287-179-4_2

there, I almost always knock, and go in and we have a chat. The thing about this is that Steve always seems pleased to see me, even though I will often have called in to ask advice, evince feelings, grumble; this does not happen only on arrival.

We often eat together, particularly when at conferences like PME – The International Group for the Psychology of Mathematics Education (http://www.igpme.org/) – to which, needless to say, Steve introduced me – and the many residential weekends he or I organise for our courses. And, of course, we often teach together. Very often our students get our names mixed up, and this makes us laugh. We write together less often, I think, than either he or I might have hoped, but we do write together. As we approach the end of our careers at our University, we are resolved to stay in touch and, at the very least, meet regularly for lunch. I don't recall a single argument.

But, of course, whilst I would like to think I have been learning from Steve all this time, I have not been studying him; I have, however, found myself doing this as I have prepared this chapter, and I should say how.

For the first time in our long association I interviewed Steve. Twice. Formally. Each for about an hour. Well, I say formally; they were meant to be formal, but the interviews could not, perhaps, but be fairly typical of our relationship, and many of our conversations: interrupted by phone calls (I think this was when Steve was finally buying a new car), me being late, colleagues knocking on his door. The first interview was semi-structured – I had an agenda: charting the social turn in Steve's writing and practice; the second, a week later, represented an hermeneutic revisiting of the first, in so far as we used the transcript of that first interview to structure our discussion. Looking back, I thought I had seen plenty of evidence over the years of the 'social' in Steve's practice, but I wanted to make sense of the idea that this might have some kind of connection to his thinking and writing. I have taken a few liberties with the transcripts of our interviews, but, after showing these to Steve, I am confident I have been true to what we meant. As usual, omissions are shown by dots (. . .) and additions by square brackets [ ]. All extracts are in italics.

These interviews were conversations between old colleagues and friends, the most subjective of research, and there would have been shared meanings, nuances which, whilst not made explicit here, were clearly there. But at this point, I remind myself that, some time back, a few years after I had been working with Steve, I realised, and I think I am right, that when Steve was writing, he was writing for an audience a large number of whom were, or would, or could become his friends. Given that sociocultural theory is at the heart of Steve's work, this book, and this chapter (and my work), I think that this is relevant. By making this point, I do not mean to detract from the value of Steve's writing, or sociocultural theorizing and research. It is, after all, the relatively recent acceptance in our community of claims that all thinking is essentially social in nature, in which Steve has played no small part, which has legitimized my writing this chapter.

So, the 'data' to which I refer here, and my reading and interpretation of our formal interview conversations, are unashamedly the result of filtering through many shared experiences, and friendship and affection, including feelings I have come to see as those I might have towards a 'big brother' (I have a big brother and I love him, but I have sometimes found myself looking to Steve as if he were a big brother): wanting to live up to him; not wanting to disappoint. I have, after all, worked with him for half of my professional life; and he has been stuck with me for nearly as large a proportion of his.

In this chapter, I have aimed to avoid too much sentimental interpretation. I have used extracts from these interviews as the bulk of the text, letting them, for the most part, do the talking (though I have added commentary in places). I have organized the extracts using a loose structure based on a rough and ready distinction between two 'phases': before Steve's PhD and after his PhD (without worrying too much about the boundary between them).

In each phase extracts from our interviews are arranged to map the social turn in Steve's writing, and the social turn as it seems to be reflected in his practice. I have aimed to produce some kind of narrative, if only to help with reading, but it is doubtful that any apparent coherence is justified, given what we know about the messiness of our human experience.

'Coming home' is a phrase that surfaced in our minds in the first interview, and I like it very much; we found ourselves using it to describe Steve's (and my) feelings about the process of moving towards sociocultural theory from the starchy, bourgeois individualism of constructivism. Coming home, I think, is a thread running through both phases.

Before the PhD – Steve's Early Teaching

What Steve says about his early teaching career suggests a trajectory – coming home – that would inevitably, I think, arc towards a socio-cultural perspective.

I think that what I have seen over the years I have been working with him has been a bit like Steve's self-explication of the theoretical basis for his practice.

Steve had encountered new maths when teaching in Israel (1970–1973). In this extract Steve refers to SMILE – (Secondary Mathematics Individualised Learning Experiment) and SMP (School Mathematics Project) and I need to set some historical and political context.

SMILE was initially developed as a series of practical activities for secondary school students by practising teachers in the 1970s (National STEM Centre 2012). I was 'brought up' from the seventies in ILEA as a SMILE teacher in London comprehensive schools. The Inner London Education Authority (ILEA) was at the single education authority for London, established by the new Greater London Council (GLC) in 1965. From 1967 to 1981 the ILEA and GLC were under the Control of the Conservative party. In 1981, with Margaret Thatcher firmly established as Prime Minister, Labour won control of GLC and ILEA. ILEA continued to be seen by many, including Thatcher, as a powerful opposition force that got in the way of Conservative plans for devolution of responsibility for education and the establishment of an educational market. The GLC was abolished in 1986, the ILEA in 1990.

The School Mathematics Project (SMP 2012) materials included individual work cards used mainly by lower secondary children. Some teachers were involved in the development of these materials, but they were much more of a ‘product’ to be bought than SMILE which was more a project to be “bought into”.

Much of my own professional development as a schoolteacher took the form of active participation in the development of SMILE within this political context, attending workshops and conferences, etc. Steve knew this, of course.

Steve: *When I took over as Head of Maths in 1975 back in England I was very keen to go . . . for the new Maths. I liked individualised learning. I did not encounter SMILE at all but I did encounter SMP the individual learning and went for that.*

[. . .] I had a kind of fairly traditional view that if you can excite [children] about mathematics, . . . if you can explain things well enough and when you realise you are not explaining well enough let them explain to each other rather than try to say it yet again; aside from that I think my teaching was probably fairly traditional. I cared a lot about kids because I had been a youth leader before then. . ..as a form teacher, I was also their Maths teacher . . . We talked about all sorts of stuff that the Church of England school would not have been happy about had they known, about drugs and about religions and all sorts of things like that. And I guess that overlapped a little into my teaching too, there was an awful lot of caring that people did not understand. I never accepted – I never really – I didn't think it through particularly, but I don't think I ever thought that ability is something that is set: anybody can learn mathematics if they get the right support and enthusiasm and interest.

Steve had encountered Wittgenstein, whose work has been a powerful influence on his thinking, in the late 70's, at the start of his PhD.

Steve: *the early Wittgenstein I had encountered but not the later Wittgenstein, of course the later Wittgenstein was the really inspiring one. I would have to say that there was nothing articulated [in my writing]. If anyone had interviewed me to say 'what is your philosophy of teaching?' well, I could have rambled on a bit but philosophy of teaching mathematics specifically, I wouldn't [have added] any coherent thought through [reference to] philosophy.*

The Constructivist

The first work of Steve's I came across was ‘Constructivism, mathematics and mathematics education’ (Lerman 1989 – This publication date is after Steve's PhD, but the work is the product of PhD and pre-PhD thinking). His stance appeared, well, constructivist. His concern in that paper had been to explore the connection between the word 'constructivism' in foundations of mathematics and constructivism in radical constructivism.

Steve: *[In that paper I was] just speculating on what the links were between the two but I mean I have not looked at that paper for years but if I remember correctly in that paper I did talk about my concerns about the private languages issue of Wittgenstein that I had encountered.*

Peter: *In my masters dissertation that I wrote .. in 1990 . . . I saw the work that you were doing as radical constructivism, would that be a reasonable?*

Steve: *Absolutely. I was somewhat confused at that stage because already in my PhD thesis I had taken a radical constructivist stance in kind of 1985 but by the time I completed it in 1986 I had encountered the later Wittgenstein. [Radical constructivism] formed a crucial part of my thesis and I had some severe doubts about [it] because of the private language issue. I just couldn't go with it. Now, in 1987 Summer I went to PME in Canada, Montréal, which goes down in history as the radical constructivist PME because Ernst Von Glasersfeld should have been giving a plenary and he was not. Instead Jeremy Kilpatrick was given the plenary to speak about radical constructivism and he was not a radical constructivist. And so the radical constructivists – Jere Confrey, and Paul Cobb, and all the others – were absolutely furious and I identified with them because it was kind of the radical younger group, if you like, and I have always, you know kind of drifted towards radical younger groups. So even though I had my reservations and even [though] we discussed those reservations I had in terms of private languages and so on, nevertheless I kind of identified with them in spirit if not in the letter of the theory. So 1987 was when I wrote that article but by the time it was out I was already. . . I had already even encountered Vygotsky which pulled me away from radical constructivism completely.*

After the PhD

Peter: *During the 10⁺-year period (1987–2000) . . . I wouldn't say that ideas [become] crystallised but you are clearly not radical constructivist. You are looking at sociocultural perspectives which you describe as a social turn, and I suppose, yes, people like me are looking to you to kind of champion the social turn at places like PME.*

Steve: *Well that certainly happened. . .. There had been a number of PME papers which were sorting out some of these ideas in 1991, 1992, 1993 those kind of years. I went to a Conference in Russia in 1993 and met Vygotsky's daughter would you believe. . . Leontiev's son [was there] – and you know Activity Theory – and it was just really exciting, and I was learning and reading; it was really an interesting, exciting time looking back on it, reading all this stuff. So the ideas had been emerging certainly and I thought that America was the heartland of radical constructivism and so to engage people in debate I had to try and get something in JRME and so I started in 1994 and it took a while because JRME bounces articles back three or four times. But it eventually went in. (Lerman 1996)*

Coming Home

About 10 years ago, partly because Steve was not always in the country, partly for my own good, I began to substitute for him, occasionally giving his lectures. As always, Steve was generous with his ideas and support, and very tolerant of my not always impressive efforts. I had been pushing Steve to talk about his huge influence on the Mathematics Education community. His response was characteristic.

Steve: *Look as you know I am a fairly modest person I think the Social Turn chapter I wrote for Jo's book (*Lerman 2000*) was a chance to bring together everything that I knew, and that is what it did for me, right. It has turned out – and I could not have known at the time – that for other people it clarified all sorts [of things] and a lot of people... I mean even [at] the Conference I went to just last week, a young woman came up to me and said 'I have to tell you that your work just transformed my thinking', and so on. And she is talking about that chapter, and it is not like I was coming up with anything particularly new there, I was putting together all sorts of things from other people and from my own work and so on, putting it all together in one, but it just seems to have been the right time for something like that and put in the right way so that people came to – people who were struggling with ideas found a lot of it there, explained for them.*

Peter: *Yes and from my point of view, in my privileged position working closely with you over that time, I suppose I would like to think that I had benefited from that both in our practice, where we talked about things, but also in terms of the development of my own understanding. I mean having to reproduce lectures that I know that you knew for example and eventually encouraging me to go and read this stuff which I think is quite difficult... Anyway...there was one other strand of questioning that had occurred to me as I prepared [for] this [interview] and maybe we should talk about which is that sense, I don't know whether you experience it, a feeling of coming to sociocultural theory as a bit like coming home.*

Steve: *Absolutely very much so. That is the best way of putting it actually.*

Peter: *And for you and me it is partly that because the point that you emphasise when you talk about Vygotsky's cultural milieu, the fact that he was coming out of the ghetto, exposed to all of these things; and of course where did you and I come from? And I do not know if you feel closer to those roots than I do – I suspect you do because of what I know about aspects of your Jewish identity – and I don't know if we want to bring that into this...*

Steve: *It is certainly relevant for me. I don't think there is any doubt about it and, you know, when planning the talks about theories of learning I love the fact that I am using a sociocultural approach in setting out how Piaget and Vygotsky came to their ideas and part of that is the recognition of Vygotsky, language, Hebrew and Russian and so on. He knew just how important they were to his understanding about the world, so it was a small step from there to sociocultural, socio-historical theories. And Piaget's world was so different, and yes it is a bit like coming home for sure.*

In our second interview, we returned to the 'coming home' metaphor. I had the feeling that we might go further, suggesting that...

Peter: *there might have been more than a feeling of coming home; perhaps we could see the appeal of the sociocultural perspective as intellectual, certainly, but maybe we should acknowledge that it is also cultural; [is there] not a kind of affective or social pull in it as well?*

Steve: *I am happy to accept that; yes, absolutely. I found an excitement about Vygotsky's ideas and part and parcel of that was his same personal history as my own; I mean, not quite of course, 50 years before, 60 years before, yeah 50 years before roughly, but nevertheless in terms of an awareness of the difference, the different world-view that one takes because of that Jewish history, that history of anti-Semitism, of Jewish humour, of that Jewish consciousness, never quite being sure that everything is all right, all those kinds of things. I think it was undoubtedly a part of the excitement of coming across his ideas.*

Peter: *Because I am not sure of how much I know [about] Piaget, I mean I certainly don't know if he ever enjoyed Jewish jokes or told them or if that is appropriate. But there's an alien-ness for me of that perspective that does not fit ... the way Vygotsky does and it is not ... just the intellectual attraction of the theory, but the fact that with it comes the acknowledgement, tacitly perhaps, of all these other factors. Have you thought of it that way before, this way I mean?*

Steve: *Yes, yes for sure, if you had asked me this question 10 years ago I would have answered in the same way.*

I think that Steve was not only turning towards the social, coming home, but also in quite a strong sense turning away from what he experienced more and more as a sterile constructivism, sterile because it ignored the role of language and mediation. A perspective that could not accommodate or even recognize the work of Wittgenstein, particularly his later work on language (Wittgenstein 1968) – hugely influential in Steve's thinking since his masters in Logic and the Philosophy of Science – was bound to be rejected at some time.

Steve realised as we talked that, whilst he had clearly articulated in his lectures to students the pull he felt towards sociocultural theory (and, I think, the perception of constructivism as 'other'), he had never actually written it down.

Steve: *I mean where I have written you know looking at constructivism and socio-cultural theory I have not drawn attention to the personal identification I feel, yeah the personal identification I feel with Vygotsky, I suppose; certainly his ideas ring true to me, not just because they seem to describe the learning experience in a much more appropriate, much richer way than the kind of constructivist theories, but also in more of a Jewish way really I think.*

What this shows is an interesting difference between Steve's theorising and practice (though the practice here is the practice of teaching about theory). What Steve, the person-in-practice-in-person (Lerman 2000) is prepared to say to his students, and the way that he says it, comes out of his thinking, theorising, and writing, but has not made it back into his writing – not yet, anyway; this aspect of the person-in-practice-in-person, I realise, has always been there for those working with Steve to see, in conversations between old friends and colleagues, and now, I hope, in this chapter.

Having made this link from Steve's thinking, theorising, and writing to his practice, I turn to a discussion of aspects of Steve's practice to conclude the chapter.

The Social Turn in Practice

Explicit evidence of the social turn is not so easy to come by in Steve's 'post-PhD' practice, but given our description of his theoretical journey as 'coming home', this should not be so surprising. After all, the distinction between theory and practice is acknowledged as, to say the least, problematic (Frade 2004, 2005). We have looked for evidence, and I think we found it.

Peter: *If now we look at your practice in those [post-PhD] years and the particular things that you were doing ... I remember ...some particular work with B.Ed students, you know the project I am talking about?*

Steve: *Yes... it was about the student teachers trying out a bit of research in their classroom, to get them engaged in teacher research because I was very interested in teacher research at the time [Lerman* 1990*]. I founded the PME working group on teacher research. So yeah it was about teacher knowledge rather than mathematical knowledge, although some of it was about teachers' knowledge [in the context of] their mathematics course. You are asking whether that kind of links with the social turn?*

Peter: *Yes I want to put it in that context because I was thinking about perspectives that were guiding you at the time... [When] you look back on [that activity], do you think of yourself doing that research as somebody ... doing it from the kind of sociocultural perspective that I would think of you as having now?*

Steve: *Yes not a fully articulated sociocultural perspective in terms of how it affected my teaching. I was busy working on the theory and the ideas, but I would not be able to say that I had, I mean looking back on it I suppose I could ... re-interpret it as playing around with development ideas, in pulling people forward in their knowledge by setting up activities and situations that would lead them to start from where they were and challenge themselves by the experiences they were encountering and the experiences of other students; but, I don't know, I don't think it was fully articulated then.*

Peter: *So the kind of contradiction to which you draw attention in your JRME paper (Lerman* 1996*) that is you cannot consistently hold both constructivist and sociocultural views, that would not have been there or you were kind of working towards that?*

Steve: *Yes I think you would have to say I was working towards that. The B.Ed [course] had gone by 1997. Yeah, I was working towards that. I had not brought together practice and the theory in an articulated way, but certainly in terms of general sociocultural theory I would say that was very much in my practice, in our practice.*

Peter: *Well I would hope so and if I recollect it – I mean at that time we are talking about I would be flattering myself to say I was struggling with the idea of sociocultural theory because I don't think I had read much, but since then I have and I can see what you were doing, at least in that sort of direction.*

In our second interview, Steve offered a useful distinction as we continued to discuss what connection there might be between his developing ideas of the social turn and his own practice.

Steve: *There are two elements, of course, just like SMK and PCK [Subject Matter Knowledge, Pedagogic Content Knowledge. See Shulman* 1986b*; Shulman* 1986a*] you know, in the work that we do as teachers in the University, working with student teachers or in-service teachers. You can talk about what influence* *your ideas have on the stuff you choose to do, . . . and you can also separately, or in a kind of overlapping way talk about the influence of those ideas on the way that you interact with students, the kind of pedagogy side; I think the first is easier to talk about than the second because, as we know, there is no direct connection between theory and, you know, teaching practices – in that under any perspective lecturing has a place, or explaining has a place, group work has a place, people struggling with ideas on their own has a place, and so on – and you just interpret it in different ways. So I could say, well I always work with groups and people can say 'well, that's constructivist, what has that got to do with Vygotsky?' and so on. So I think it is easier to talk about the first than the second I don't think there is any doubt that, you know when you are looking [at current issues in research in mathematics education]. . ., as we did in the early days of working on the Masters course when it was an MSc in Maths Ed, . . .what one talks about are the things one is interested in. If I talk about research methods I talk about research methods that are related to the kind of research you would do from a sociocultural perspective, and all the learning theory stuff that we do is meant to be an overview; but I don't hide the fact that I have my own preference, my own point of view; . . .in fact give a kind of socio-cultural interpretation of learning theories. . . . what I am saying is, you know, it is not difficult for me to say how I conceive of my pedagogy in sociocultural terms, [but] what I am suggesting is that anybody reading it could say 'well, why is it sociocultural,?' That is all I mean.*

Peter: *. . . [you] are not just saying, presumably, 'Oh, I will teach in this way because it is a good idea', you are saying this is how people learn. . .*

Steve: *Yeah, I suppose what I would answer is [that] I can describe what I do as working in the zone of proximal development, or I can describe what I do in terms of mediation, and I think about what I plan to do in those terms as well, whereas in earlier more naive days, I might have said 'well, I am going to let people construct these ideas for themselves'.*

It seems to me that Steve's practice has always been at least consistent with sociocultural perspectives; if there is any power in the metaphor of 'coming home', then we might well expect to see tacit evidence of this in his practice even before its articulation; perhaps we might best see the relationship between Steve's practice and his theorising and research in terms of a process of construal, making sense of dispositions to work with students in certain ways that owe something significant to a cultural milieu, happily less threatening, but not so far removed from that into which Vygotsky was born.

Looking back at the development of Education courses within our University, Steve's evolving perspective has clearly informed his own contribution to this activity, but it also, to use the language of activity theory, has become a powerful mediating artefact within that activity. It directly influenced those of his colleagues who, like me, were working away at course development; just as formatively as the funding opportunities and policy drivers coming from government agencies, it shifted the object of our activity to include, quite explicitly, ways of working with course participants made possible, I think, only by the adoption of the kind of sociocultural perspective Steve has been articulating. I think there is more evidence of this in our second interview.

Steve: *... like our school-based MA, the EdD [I set up six years ago] was ... kind of a policy development. Other universities were offering EdDs. It looked like a great opportunity to get people working together on higher level thinking, doctoral level thinking which sounded very exciting, and it was both an expansion of courses that the Department offers and an opportunity for a new kind of teaching which was really a continuation of the Masters programme, I suppose, at a higher level. So it was kind of policy and new initiative and income and development and so on, portfolio development, that drove it; but I think, having just said what I have just said, ... it is clear that part of our thinking also was [to do with] what we have done very successfully on the Masters course: people in practice are talking about their practice and researching their practice and thinking about what researching your practice might actually mean and accessing what literature there is around in a much deeper way, but very much the same kind of thing; people working in groups, sharing their experiences. Social turn, yeah I mean it is part and parcel of it, I don't know.*

Peter: *I suppose we might ask if had you not taken the social turn you could imagine the EdD developing [and in] that way.*

Steve: *An EdD developing I am sure. I think it is highly likely because, as I say, it was driven as much by policy, local and national policy, as anything else ... but perhaps it would have been harder to conceive of the value of a course structured such that people share their experience with each other and pull each other's learning along .... I guess I would not have conceived of it in the same sort of way.*

Steve was being typically modest here. What those of us working with him have been fortunate enough to have presented to us – and may sometimes have taken for granted – as part of the intellectual and collegial air we breathe, infused with a richness and depth of understanding of 'the social' in teaching and learning, others have had to take from their reading of his papers (though Steve has made this easy for them, of course). Steve's chapter in Jo Boaler's book (Lerman 2000) has clearly provided pivotal insights for many around the world; it did for us too, but less dramatically. Steve's unit of analysis 'person-in-practice-in-person' (ibid, p. 38) is helpful here; closeness to Steve has, I hope, allowed us every day to brush up against and learn from this lovely 'person-in-practice-in-person'.

References

Frade, C. (2004). The tacit-explicit dynamic in learning processes. In M. J. Høines & A. B. Fuglestad (Eds.), *Proceedings of the 28th conference of the International Group for the Psychology of Education* (pp. 407–414). Bergen: Bergen University.

Frade, C. (2005). The tacit-explicit nature of students' knowledge: A case study on area measurement. In *Proceedings of the 29th conference of the International Group for the Psychology of Education* (pp. 327–334). Melbourne: University of Melbourne.

Lerman, S. (1989). Constructivism, mathematics and mathematics education. *Educational Studies in Mathematics, 20*(3), 211–223.

Lerman, S. (1990). The role of research in the practice of mathematics education. *For the Learning of Mathematics, 10*(2), 25–28.

Lerman, S. (1996). Intersubjectivity in mathematics learning: A challenge to the radical constructivist paradigm? *Journal for Research in Mathematics Education, 27*(2), 133–150.
Lerman, S. (2000). The social turn in mathematics education research. In J. Boaler (Ed.), *Multiple perspectives on mathematics teaching and learning*. Westport: Ablex Publishing.
National STEM Centre. (2012). http://www.nationalstemcentre.org.uk/elibrary/collection/44/smile. Consulted September, 2012.
Shulman, L. (1986a). Paradigms and research programs in the study of teaching: A contemporary perspective. In M. Wittrock (Ed.), *Handbook of research on teaching* (pp. 3–36). New York: Macmillan.
Shulman, L. (1986b). Those who understand: Knowledge growth in teaching. *Educational Researcher, 15*(2), 4–14.
SMP. (2012). http://www.smpmaths.org.uk/. Consulted September, 2012.
Wittgenstein, L. (1968). *Philosophical investigations* (3rd ed., p. 271). Oxford: Basil Blackwell.

Chapter 3
Steve Lerman: The Man and His Work

Gilah Leder

> *I was a secondary school teacher of mathematics in various schools in London and abroad, including 5 years as a Head of Mathematics, before writing a PhD and moving into Higher Education*
> (Steve Lerman - *Staff profile, London South Bank University*)

In broad terms, Steve's biographical note mirrors the pathway and entry into academia of many members of the mathematics education research community. Burrowing a little beneath the surface, and learning more about Steve, the man and his work, are the overriding aims of this chapter. Throughout, I draw heavily on Steve's own voice, the perspectives of some of Steve's colleagues, and the copious amount of written materials publicly available.

I considered a number of different models for writing this chapter. Ultimately I was strongly influenced by the approach used by *Inside the Academy*. "Inspired by the Emmy award winning *Inside the Actors Studio*, *Inside the Academy* honors the personal and professional achievements of exemplary scholars in the field of education" (http://insidetheacademy.asu.edu/). The biographical interviews touch on personal as well as professional issues and include "candid reflections from family and friends". Core information for this chapter was gained via an extensive face-to-face Skype interview conducted with Steve Lerman in the early hours (London time)/late afternoon (Melbourne time) on May 21, 2012. Various colleagues of Steve's also provided information. Selected excerpts from these contributions are included in the chapter.

The Lerman Family

Steve was born in the UK, the oldest of three children. It was a close knit family, a description that applies equally well to the current generation. Thanks to some time spent as a stay-at-home father when his two daughters were quite young, Steve is particularly close to them.

G. Leder (✉)
Monash University, Melbourne, VIC, Australia
e-mail: gilah.leder@monash.edu

P. Gates, R. Jorgensen (Zevenbergen) (eds.), *Shifts in the Field of Mathematics Education*, DOI 10.1007/978-981-287-179-4_3

> We have such lovely common memories of me taking them to their ballet classes. They were forever doing competitions and shows. I think back to one of these, where I was busy plaiting the hair of one of my daughters into a French plait as they call it, and helping the other one do her makeup when they said "dad look behind". And there was this whole group of the mothers just astonished that this father was doing exactly the same as they were. We can think back to lots of these wonderful experiences and memories, and that has made us very close.

Now grown up, one daughter is a primary school teacher who may well be moving towards an academic position. The other is a well-travelled senior executive for a large enterprise. The "role swap" alluded to above was not only important for Steve's career development but also enabled Steve's wife to follow in her father's footsteps and pursue her own career. "I think she has enjoyed being a business woman", Steve reflects.

Becoming a Mathematics Teacher

Career choice can be serendipitous or be subtly influenced by a number of factors. In Steve's case, three features stand out. He enjoyed doing mathematics and was good at it. As a teenager he had joined the youth group Habonim, soon took on a leadership role, and realized how much he enjoyed working with young people. And third, but certainly not least, was his close relationship with, and affection for, cousin Gershon (Gerry) Rosen:

> Gerry Rosen was a kind of big influence because his mother and my mother were twins and then he grew up next door to me and he was a couple of years older than me, so he was like a big brother, not really a cousin, much more a big brother. In fact we looked quite similar. He went into teaching mathematics and I had that consciousness that that's what I was supposed to do.... So there was something of a family influence if you like. I reflected on it in later years and realised that was quite an influence.... So putting everything together, by the time I went to university to do my degree in mathematics, I already had it in mind that really all I wanted to do was to be a mathematics teacher.

Nevertheless, the route to teaching took some time. As Steve recalls:

> I started university in 1963 and began teaching in 1966 but then soon stopped that and began working for (the youth group) Habonim as a youth leader. It was only when Beryl (Steve's wife) and I went (in 1970) to a Kibbutz in Israel that I picked up the teaching of mathematics. I had to learn Hebrew first of course, but once I'd learnt Hebrew I started teaching in the local Kibbutz high school, teaching in Hebrew, and that launched me on the mathematics educational teaching and research career.

The Lermans spent two and a half years on the Kibbutz. At first Steve's Hebrew was not good enough for him to do any teaching but that changed over time. "That last year", he recalls, "was very enjoyable and very successful and then we left and came back to England and I picked up teaching again". At that stage the possibility of a research career had not been considered, even for a moment. "I imagined myself being a teacher all my life". (See Peter Winbourne's chapter on Steve's teaching).

By 1975 Steve had become the Head of Mathematics in a secondary school. It was a position he relished: "I really enjoyed it, it was great. I enjoyed co-leading a team and working with people on pedagogy and on curriculum. But (by 1979) the question began to arise after on-and-off teaching for many years, what am I going to do next?" Administration held no appeal (an interesting confession, given Steve's many sustained leadership contributions to the profession); nor did headship or deputy headship seem an attractive prospect. "So", Steve recalls, "I was wondering what to do next".

Moving Away from Mathematics Teaching

What ultimately makes a good researcher? Ingredients certainly include an ability to learn from experience, to ask questions, to reflect, to go beyond one's comfort zones and embark on ventures with an element of risk. These features are unquestionably embedded in Steve's account of moving away from teaching into higher education – the latter initially as a student.

> I guess lots of questions had arisen for me in being head of mathematics. We opted for individualised learning materials, and there were other kinds of decisions that I'd made and I realised I didn't have any basis for making these decisions. Why would individualised learning be better than individual learning – the way the rest of the school worked in every other subject? And so I thought, well I enjoyed the theory part of my teacher training course that I did, the PGCE course. I particularly liked philosophy of education as it happens, but I thought maybe I should just do some more studying. I had a Masters Degree in Logic and Philosophy of Science which I'd done part time, and so study was in many ways part of my life. You know it's a technical Jewish thing anyway to spend your life studying. I thought about doing a PhD and applied in a few places and got accepted at Cambridge, initially to do philosophy, just philosophy of education.

That first year of doing his PhD, under the supervision of Paul Hirst – already a leading philosopher – was a particularly enjoyable time.

Changing directions mid-career is rarely unproblematic. Nor was it so in Steve's case. He did not have a grant to work further on his doctoral studies, doing a PhD part time at the University of Cambridge was not an option, and tuition fees cost "a fortune". As mentioned earlier, he and Beryl had swapped roles. While she went out to work, Steve looked after their two daughters and combined this with his new role as a student – a role he was not ready to give up.

Transferring universities and supervisors was not particularly straightforward. While considering various options there was also Paul Hirst's enticing advice to consider and explore: "I think you should probably pick up some philosophical work in mathematics education rather than pure philosophy. You should bring in your teaching of mathematics", he memorably told Steve.

> Eventually somebody said "why don't you try King's?" I had a chat with Margaret Brown and there were scholarships available and I got one of them, the SSRC Scholarship as they were called then, and began seriously in 1982 at the same time as Richard Noss.
>
> With the change of institutions and supervisors came a subtle change in PhD topic. "It was in teachers' beliefs, but it was very much influenced by philosophical approaches to mathematics".
>
> Steve's time as doctoral student at the Chelsea Centre coincided with the period that many subsequently influential and productive mathematics educators worked there – for example, Kath Hart, David Johnson, and Margaret Brown. They were among the group who moved to the King's site when Chelsea College was absorbed into King's College London – around 1983.

A part time position at the Institute of Education, University of London became available at a, for Steve, convenient time. He applied, successfully, for the position of PGCE tutor, a job he held for 2 years while completing his PhD. It was an arrangement "that fitted in well with still looking after the children". Around the time he completed his PhD, a temporary full-time job came up at the Institute which he filled, for 1 year, 1987. From there he moved to a position at London South Bank University where he has remained since then and, in 1998, was promoted to Professor of Mathematics Education.

Academic Life

How It All Began

Looking back on those early years of transition from secondary school mathematics teacher to working in a tertiary institution, two forces stand out.

> First of all in those early days Paul Ernest and I became very good friends because of our common interest in philosophy of mathematics. I think that helped a lot – helped both of us a lot because we talked an awful lot and argued and discussed and read and so on.

Paul describes those early, and subsequent, interactions as follows:

> I read Steve's PhD thesis in 1987/8 and it was part of the inspiration for my book Philosophy of Mathematics Education (Ernest 1991). Steve suggested the title to me after reading the transcript in 1989. My intended title was Mathematics Teaching Philosophies. We met, with several others, every year at PME from 1986–1995 and also between conferences. I solicited papers and chapters from Steve in several of my edited books. (Lerman 1989, 1994)
>
> Back in London Steve and Marilyn Nickson had instituted a reading/sharing group on the Social Aspects of Mathematics Education. That was very exciting and formative for me – lots of good people were members. The group, which ran from about 1987/8 until early 1990s and met at least three times per year, included Celia Hoyles, Richard Noss, Jeff Evans, Steve, Marilyn, Paul Dowling, me, Candia (Morgan), and lots of others. We produced several papers which we typically circulated in advance. A book of papers came out of it – but much more also...

The rich source of intellectual offerings available at PME [The International Group for the Psychology of Mathematics Education] and the opportunities for making new contacts at the annual conferences of the organization also proved invaluable. What still stands out for Steve was attending his first PME conference and the horizons that broadened.

> . . .My first PME, in 1984 in Sydney, (Australia). . .. That was very, very significant because I met all sorts of people – you know the people whose articles I'd been reading – something almost everybody says when they first attend PME – books I'd been reading, here were the very people! I got involved quite quickly in Discussion groups and Working groups and presenting papers. So PME was a huge influence in terms of pushing ideas forward, getting me to write. I had no mentor at all either at South Bank or continued mentoring from my supervisors at King's. I finished my PhD and waved goodbye and that was it. I was really floundering, and if it wasn't for PME, with that structure of Research Reports, Working group and publications in Working group, I think it would have been much harder to have got as involved as I did in researching and writing.

What directions were spawned by his reading, interactions, and debates with colleagues? Only the bold outlines are described in this chapter. More detailed information is found throughout the book, in many other of its chapters. The beginning of his new intellectual journey is captured here in Steve's own words:

> In my thesis I was quite critical of constructivism whilst feeling that that was a fruitful direction. But after completing my PhD work that influence of language became something I became really interested in and took me away from constructivism and I began reading Vygotsky's work and somehow I found it inspirational. I don't know how much I was also influenced by the fact that he was Jewish and writing about the influence of language. It occurred to me that if you think about Yiddish and you think about Hebrew, and you think about Russian, you realise that these are not just different languages like – I mean Piaget was multi-lingual I'm sure, being Swiss he no doubt knew German, French, and Italian, and English too, but they're much more similar than languages like Yiddish and Hebrew and Russian. And Yiddish, Hebrew and Russian carry world views with them. Carry histories with them, and I suppose it seemed to me, as a fellow Jew, just how obvious it was, that language forms you. That language is culturally based and shapes you as a cultural people in the richest sense of that - whether gender or social class - but also religions, ethnicity, histories. So I found Vygotsky's work very powerful and quite soon began, I suppose, the kind of drift towards a discursive shift, or socio-cultural shift, which I've written about and researched extensively over the past 15 or so years.

Favourite Works

Like many researchers, Steve has his favourite projects and publications. Only a few pieces are mentioned here. [A more detailed and in-depth analysis of Steve's work is again found in other chapters of the book.]

> First of all I'd have to say the JRME article published in 1996 (Lerman 1996) which was a critique of constructivism and a proposal that a socio-cultural, a full socio-cultural turn to the socio-cultural theories was much richer for our understanding of children's learning and of teaching. I was particularly proud of that because North America was where radical constructivism was strongest, is still strongest, and I felt it would be all very well to try and

get a paper like that published in Educational Studies in Mathematics, but among its readers there would be more people who would naturally find that closer to their own thinking. I wanted to get an article into the heart of the camp of constructivism. As with all JRME articles it took some time with the editing process but eventually it was accepted in 1996. And I thought this is great, you know, I'm now going to be debating with the whole of the American mathematics education society, and then of course I never heard a word. For a long time I had no idea whether anyone read the article or if it had made any difference to anything or anybody. But two years later, at a PME meeting I came across somebody who had just finished his doctorate at Georgia. This was at the time when Georgia was the heart of Les Steffe, Ernst Von Glasersfeld, and Paul Cobb country. I remember saying to him "you know I wrote that article and I don't know if anybody ever read it." "You're joking", he said, "that is an essential theme of all the doctoral work at Georgia, it's THE article that we have to tear apart in order to get approval to the next stage of our PhDs". So that made me feel somewhat happier. I still think about that conversation.

The chapter written for Jo Boaler's book, published in 2000, is cited as another favourite (Lerman 2000). It is a publication that has been well received and widely quoted. More than a decade after it was published Steve reflects that: "It seems to have been significant for a lot of people, presumably because it synthesized a lot of socio-cultural work and gave people many directions to look further. I am particularly proud of that chapter."

The project by the Economic and Social Research Council [ESRC], "UK's largest funder of research on economic and social issues" must also be mentioned here. An important aim of the work which spanned a period of 12 years was to identify theories of learning adopted by researchers. Steve, who was the lead researcher on the project, considers it possibly his most substantial piece of research.

We drew on theory and worked hard on the theory. We had Basil Bernstein's support and advice....We did a lot of empirical work. The only chance I had to do such rich work was for my PhD and the ESRC project. The publications that came out of there were, also for me, quite important.

Anna Tsatsaroni was a close collaborator in this work. She, Steve notes, "helped me an awful lot to get into the sociological literature, to understand sociological concepts and Basil Bernstein's work in particular" all elements which have been integral to Steve's research over the past two decades. Anna reveals:

My own perspective originates outside mathematics education, precisely the sociology of education. My interest in the way school knowledge is organised and its probable effects on students (and teachers) meant that I have focused my research on curriculum, pedagogy and assessment practices. Furthermore, Mathematics (and science) was ideal since mathematical knowledge was/is considered to be epistemologically 'hard' forms of knowledge, its objectivity difficult to challenge and therefore school knowledge as a social construction difficult to develop as an argument.

Steve was ready to engage with these sorts of argument and he was seeking to support the opening and firm grounding of this perspective. The sociological theories and in particular Bernstein's sociology of knowledge therefore came naturally on the forefront of our conversations (Morgan et al. 2002).... The question posed at the first Conference on Mathematics, Education and Society (MEAS1) (Lerman and Tsatsaroni 1998): "why children fail and what mathematics education studies can do about it" was the beginning of our long term collaboration. There were many participants in that conference who did not know Bernstein's work and some who had some knowledge of it, but after the plenary were encouraged to read more about his work.... In exchanges with Steve, from then onwards, and in presentations in conferences the issue of global policies and effects on (mathematics) education research was indeed thematised.

Other Collaborations and Contributions

Over the years there have been collaborations with many other researchers – some on a daily basis within London South Bank University; others with colleagues in the wider mathematics education research community. Peter Winbourne is among the former. Working closely with Steve over many years, he told me, has been

> ... a real privilege, an opportunity to work with someone who is very wise and steeped in scholarly activities. He has always encouraged me, included me in his work, introduced me to people all around the world and undoubtedly stretched and challenged me in that I did not want to disappoint him ... This began from the moment I started work at South Bank. I know that his work is read widely, but somehow you often get the feeling that he is writing for his friends, that the writing is part of a continuing dialogue in which he encourages others to engage. This accessible scholarship also comes out in his teaching; the warmth and respect for the work and views of others.

Various collaborations with colleagues beyond the walls of South Bank University are listed by Steve. Among those mentioned are the classroom studies with Leonie Burton – a project which was never finished because of her untimely death. In the 1990s there was some work with David Clarke in Australia.

> He (David) invited me, and others, over for work on a classroom project. He was getting some really rich data from science and mathematics classrooms. I was very pleased to have been invited into that, and I'm proud of the paper that I produced. It was a study of two boys working on mathematics, algebraic cancelling actually. I gave an account of what they were doing entirely in Zone Proximal Development social influences analysis (Lerman 2001), avoiding the need to write about cognition in a deep psychological way.

More recently there was the now completed EU funded ABC-Maths project, concerned with "big ideas" in mathematics and in mathematics pedagogy.

> This project was both interesting and very fruitful. For us in the UK it was based on a course on teaching advanced mathematics that had been running for five or six years. The course is still running, and so – even though the project itself has finished – we are continuing to work with the teachers on these big ideas and researching what effect it has on their teaching.

There have been collaborations with Robyn Jorgensen (previously known as Robyn Zevenbergen). Two projects stand out. In Robyn's words:

> The first, funded by the Australian Research Council, operated between 2003 and 2005. The study explored the ways in which teachers were using ICTs to enhance (or not) aspects of mathematics education. The study was conducted in the primary years of schooling and involved a sample of schools that encompassed the diversity found in Australia – including schools in affluent areas, socially and economically- disadvantaged schools and schools in rural areas of the country. Over the period of the study, the implementation of interactive

whiteboards was becoming common and so the study was expanded to include this area of ICT use in the classroom.

The second, and still ongoing study, involved not only Steve and Robyn, but also Jo Boaler and Peter Sullivan and is concerned with mathematics teaching and learning in remote Aboriginal communities. The larger team has been involved in several other, USA-based projects.

There have been, and still are, sustained periods of teamwork with Jill Adler. At times such collaborations have consisted of contributing to the graduate research program in mathematics education at the University of Witwatersrand. This has involved helping doctoral students not only at their home university but also when they have come to study in the UK, generally for a period of between 1 and 6 months. Working together with Jill and her colleagues on both smaller and large research projects has been another, and on-going element. In recognition of the value of his extensive and diverse contributions, the university appointed Steve as a visiting professor.

In Sweden, too, Steve's role in developing and expanding the country's mathematics education research community has been recognized. Several years ago he received the Sven Pedersen Award.

Steve has relished his collaborations with colleagues. "Being part of other people's projects has been very enjoyable and fruitful. But", he noted wryly, "an awful lot of work has been me and my study, reading, writing, doing small scale work."

Steve's students have followed many different pathways, with only a few carving out a career as mathematics education researchers. Anne Watson is among these and a collaborative project scheduled to start later this year is now on the horizon. Looking back, Steve remarks that working collaboratively with former students is an area that has "not been as plentiful or fruitful as it might have been. Regrettably there were limited opportunities for this at South Bank".

Leadership and Administration

Within the University

Despite volunteering that administration was far from his favourite activity, Steve served at different, and for varying lengths of times, as Course Director, Head of Department, Deputy Director and Director of the Centre for Mathematics - the fairly small research centre at London South Bank University. Although only some of the planned activities came to fruition, it provided more welcome opportunities for collaborative work. Candia Morgan who was part of the Centre of Mathematics team during her time at London South Bank University described Steve's influence as follows:

Steve has made a massive and generous contribution to my personal professional development. I shared an office with him at the very start of my academic career and he took me under his wing, always supportive and encouraging but also, crucially, involving me in opportunities to do research with him even after I left South Bank University.

Although my own research focus is not close to Steve's main interests he has always shown interest, appreciation and engagement with ideas. I think this is a hallmark of his contribution to mathematics education research – a breadth of knowledge and engagement with a range of issues, theoretical perspectives and methodological approaches. For me, his most significant substantive contribution is his advocacy of and contribution to the development of perspectives on mathematics education that take account of the social. His research has made an original contribution to thinking in this area but has also succeeded in opening up the field as a whole to accepting theoretical and methodological approaches that draw from sociology, linguistics and other disciplines that focus on social concerns.

Looking at Steve's strong and successful leadership profile, it is clear that he has been both selective and discerning in the administrative roles he has elected to take on. What made certain roles seem both attractive and worth doing?

Within the Wider Mathematics Education Research Community

Earlier in my career I could have headed towards deputy head. Back in Habonim days I became the General Secretary of British Habonim and it was pretty big at the time. Twelve Youth centres and lots of movement workers, and a fair budget. I enjoyed doing that. And although I did not stay in that position for a long time, I enjoyed the work and responsibilities. So it's not that I find administration difficult. Rather, it's that I'm not excited by that kind of work - except for the things that I'm interested in. But PME just came along and that was great. It proved a hard job but it was infinitely rewarding. I felt it was an honour, and it was always a challenge because you feel you've got this history in your hands and you could drop it and it can smash if you're not careful. Although I'm sure others would never have let that happened, but still, you feel that responsibility.

The fact that the group continued and was successful, that the conferences were successful and PME's membership continued to build up, that was great and it was great doing the job of President. It was strange that just at that time I also became a Chair of the British Society for Research and Learning Mathematics [BSRLM]. The two happened almost simultaneously. And again I enjoyed that position. It was a particularly crucial time for BSRLM which has, more or less, the same length of history as PME.

Steve took over as chair of BSRLM at a critical time for the organization.

There was no structure and no system of elections or anything like that. Whilst everybody enjoyed that in a way, it's dangerous, because it then depends so much on casual volunteering and it can begin to fall apart - as was happening. So I used the PME experience to build structures of elections and committees and a system of meetings. . . . That process has continued and the organization has grown from strength to strength. It was an honour to be elected, and a responsibility. Fortunately it worked out very well.

The Future

At the time of writing this chapter, in Australia and thus well away from the UK, the university scene in the latter country seems to shout contradictory messages of expansion and contraction. According to Steve, too, "some places have the courage to expand and others are just scared of the future". What, more broadly, is Steve's vision for the future?

> It's an interesting question. What do we know in our field? What have we, mathematics educators, actually found out? Suppose we all sat around and spent three or four days trying to make a list of the research findings that we all considered incontrovertible, that we all absolutely agreed with and used in all our teaching. What would happen? How many things would we write down?
>
> I came into teacher training knowing what being a good teacher was, because I'd been a good teacher, I think. Since then there have been a series of more and more questions. I used to joke that if I went back into a classroom I'd stand there and look at the kids and I'd think constructivism, ZPD, group work, individual work, help what do I do? Of course you don't. If you walk into a classroom you switch back to what you always used to do.
>
> We've raised more questions than we've provided answers. I don't think that's a bad thing though. I know there are some who say "with all these theories around we can't even talk to each other." But I think in some ways it's very fruitful because people can work in different ways and different directions and if we think about teaching as helping people to find how they can be as good a teacher as they can be and continue to learn about teaching. That's better than saying "I know what good teaching is and I'm going to help train you to be like that".
>
> So, I'd like to think that in the future there would be room for teachers and researchers to continue to experiment, continue to learn, continue to find different directions, more theories I suppose, and more questioning. I think that's the healthy thing. I don't know whether we'll have the freedom to do so because our governments are being more and more prescriptive, telling teachers what to do, inspecting them to make sure they're doing it. So there'll be that kind of tension I think. And of course there is less money for research grants than there once was. So it's an uncertain future, I think, but also exciting in many ways because of this opening up, this greater diversity and multiplicity of theories. But that's something I'd like to discuss rather than pronounce on.

A Final Word

As is readily discernible from the contents above, research for this chapter involved contacting a number of Steve's colleagues. Instead of quoting, inevitably selectively, from the many responses I received I want to conclude the chapter with the note sent by Ted Eisenberg, who most evocatively captured many of the elements put forward by others.

> **Stephen Lerman: researcher, teacher and friend**
>
> Steve's main academic interest is in investigating the impact various theories of learning have on teaching, and he has written many papers in this area. I will let others assess the impact of this work, but suffice it to say that one does not become the President of the international organization PME, or the Director of London South Bank University's Center for Mathematics Education or the Deputy Director of their Institute for Research in

Education, nor a main researcher for the EU-funded ABC-Maths program, or be selected as a co-investigator for an Australian Research Council funded project that deals with remote and Indigenous schools in Western Australia, if one hasn't a long and impressive track record in the field. Steve certainly has such an impressive record and the above listings are just some of his recent activities. But let me address a few sides of Steve to which I was exposed when he graciously shared his office with me during a sabbatical year that I spent in London.

Steve is an incredibly hard worker, arriving at the office early and staying late. He is also incredibly efficient in answering phone calls and simultaneously typing, what seemed to me to be a mile a minute, on his computer. He would often "eat on the run" to take advantage of every minute of the working day. All of these things about him impressed me, but what impressed me most was the patience he had in helping a struggling student. There are many examples of this but let me mention just one.

A student from a non-English speaking country stopped by Steve's office one day to consult with him. Listening to their conversation I soon learned that this student was enrolled in South Bank's graduate program and that she was having some trouble in understanding a homework assignment Steve had given to her that dealt with the Tower of Hanoi puzzle. There are many questions that one can ask that stem from this puzzle, but it soon became apparent to me (and to Steve) that this graduate student in mathematics education was encountering the puzzle for the first time, and worse, that she seemed not to know mathematics at all. My initial reaction was one of being aghast, because if I couldn't recall having ever before encountering a graduate student in our discipline who wasn't familiar with the Tower of Hanoi puzzle. But here was one speaking to Steve and my initial reaction was that he should immediately kick her out of the program. (As an aside, let me mention that I always felt that the atmosphere in Steve's office was workaholic; colleagues did not stop by to chit-chat. Everyone seemed to me to be in a hurry; with everyone multi-tasking to the hilt. I rationalized all of this by reminding myself that this was London where everyone is on the go and in a hurry.) Anyway, when Steve realized that this student had some humongous holes in her background he became unbelievably gentle and devoted his complete attention to her. He did not get upset nor did he dismiss her. He showed her respect and he was patient with her, unbelievably patient. I recall thinking to myself that I would not let such a person in our graduate program here in Israel, but that is easy for me to say because I wasn't dealing with her, Steve was and she was already in their graduate program. Anyway, Steve worked with her and at the end of the session she seemed to understand the puzzle and how to go about her assignment. I don't know what ever happened to this student but I learned something from Steve that day; that initial impressions of someone's intellectual capabilities are often wrong and that everyone, absolutely everyone, has the ability to surprise you. Steve taught me that day to hold off making initial impressions. It is a lesson I hope that I will never forget.

Steve is a caring individual who has contributed to so many of us in ways of which I am sure he isn't even aware. Yes, he is a researcher with an international reputation and his contributions to our field have been immense, but more importantly, he is a teacher, and he is my friend.

References

Ernest, P. (1991). *Philosophy of mathematics education*. London: Falmer Press.

Lerman, S. (1989). Investigations, where to now? In P. Ernest (Ed.), *Mathematics teaching: The state of the art* (pp. 73–80). London: Falmer Press.

Lerman, S. (1994). Articulating theories of mathematics learning. In P. Ernest (Ed.), *Constructing mathematical knowledge: Epistemology and mathematics education* (pp. 41–49). London: Falmer Press.

Lerman, S. (1996). Intersubjectivity in mathematics learning: A challenge to the radical constructivist paradigm. *Journal for Research in Mathematics Education, 27*(2), 133–150.

Lerman, S. (2000). The social turn in mathematics education research. In J. Boaler (Ed.), *Multiple perspectives on mathematics teaching and learning* (pp. 19–44). Westport: Ablex.

Lerman, S. (2001). Accounting for accounts of learning mathematics: Reading the ZPD in videos and transcripts. In D. Clarke (Ed.), *Perspectives on practice and meaning in mathematics and science classrooms* (pp. 53–74). Dordrecht: Kluwer.

Lerman, S., & Tsatsaroni, A. (1998). *Why children fail and what the field of mathematics education can do about it: The role of sociology.* Plenary paper presented at the First Mathematics Education and Society Conference (MEAS1). Retrieved from www.nottingham.ac.uk/csme/meas/conf.html

Morgan, C., Tsatsaroni, A., & Lerman, S. (2002). Mathematics teachers' positions and practices in discourses of assessment. *British Journal of Sociology of Education, 23*(3), 445–461.

Part II
Steve Within the Field

Chapter 4
Issues of Equity and Justice in the Construction of Steve Lerman

Peter Gates

Background

In many chapters of this book, we read of specific contributions Steve has made to various branches of our field, each written by colleagues who have known Steve for many years. Although I too have known Steve personally for almost 30 years, because of his being so self-effacing, one rarely gets to know the man too deeply. In order to write this chapter I met with him and talked – enjoyably at length – specifically to look at his political background and orientation. The quotes I intersperse in this chapter are taken verbatim from that conversation.

My focus in his chapter is Steve's politics and his commitment and work in and around equity and social justice. . . .That might sound easy – but it is not. In the entry on Equity of the Springer on-line Encyclopaedia of Mathematics Education, I argued that it was not possible to give a dictionary definition for equity – indeed I go so far as to say the same for social justice.

> With both "equity" and "access," that's not possible. Each of these terms is politically loaded and reflects political and ideological dispositions both in the pedagogical arena of the classroom and in the intellectual arena of the academy. (Gates 2014)

One problem for defining equity is that it is assumed to be a universal good; surely everyone wants equity? Actually that's far from the case. There is little agreement on how we define and more importantly operationalize the term. Equity is not a key driving force for those who sit on the political right; there, meritocracy and individual endeavour are markers of a democratic society, providing a way out of poverty for those who work hard. For those on the political left, the economic superstructure itself, and the education system which serves that system, hides

P. Gates (✉)
School of Education, University of Nottingham, Wollaton Road, Jubilee Campus, NG8 1BB Nottingham, UK
e-mail: peter.gates@nottingham.ac.uk

P. Gates, R. Jorgensen (Zevenbergen) (eds.), *Shifts in the Field of Mathematics Education*, DOI 10.1007/978-981-287-179-4_4

structural inequality and merely perpetuates that structural inequality based on accumulated wealth. For the left, equity itself is a key feature of a democratic society; for the right it is an impediment.

A key question for me is "*how does your work change the position for the most disadvantaged in society for the better and at the expense of the most advantaged*?" I asked Steve this very question......but will hold onto his response until the end of this chapter.

I see there being two approaches to equity. The first is *illumination*, where we point to and examine the existence of differential outcomes. The second is *emancipation* where the focus is more directly on doing something to redress the inequity through a grasp of the deeper forces that bring it about. Robyn Jorgensen and I argued in a recent paper (Gates and Jorgensen 2009), that whilst no-one can publicly admit these days that they are not committed to or interested in social justice, there are in reality different positions which individuals take that give away their real commitment: moderate, liberal and radical. Moderate forms of social justice focus on fairness and equity but presume the continuance of the political status quo and do not explicitly recognise or relate to structural inequalities in society.

> Work in this tradition might typically focus on classroom relationships, language, and assessment. This form of social justice might easily be signed up to or even hijacked by neoconservatives, who recognise inequality but who want to avoid a questioning of the difficult social conditions which bring it about and the potentially threatening challenges required to bring about change. (Gates and Jorgensen 2009, p. 166)

A second form of social justice does recognise structural inequalities but with an acceptance that classrooms can be made more just within the existing structures. "*Liberal forms of social justice see relationships as a key feature in classroom interaction, and may go further and see a classroom as the main social organ*" (p. 167). The third and radical form of social justice recognises structural inequality and seeks to redress the ways in which inequality is built into existing practices. Such a position assumes that it is only by changing these structures that greater access to, and success with, mathematics is made possible for those groups of students who have been excluded from participating in mathematics.

So, I guess I have to ask where Steve lies on this classification. I am hoping to demonstrate that although Steve's writing does not always address the ways in which structural factors bring about inequity in mathematics education, nevertheless one can see how this is the direction Steve's work is pointing, and is the window he opens through which others can jump.

Meeting Steve

Many colleagues in this book have described how they first encountered Steve. The first moment, event or situation when I first met Steve is now lost in the mists of time, but is likely to have been around the time of Steve's first PME (8) in Australia in 1984 and my first PME (10) in London in 1986. Both of us were in the same phase of our lives of moving from Head of Mathematics in a school, to academic, researcher, and higher education worker. They were exciting times. Many of us were then the angry young men and women of mathematics education, no doubt aggravating the old guard. Some of us (me included) continue to be angry and aggravating. Steve could never be described as either of those things. On the other hand, he has always appeared to be able to remain calm and reflective. Yet, like very many of us, he has a history which places him firmly on the left.

> As a young boy I was bought up in London, between Hendon and Golders Green. I think my father voted Labour, I don't know about my mother. As a young teenager, they sent me to a Zionist socialist club where there was an injection about Jewish history and socialism. They were both Zionists but really more wanted me to mix with other Jewish children. As a group we got interested in Jewish writers, reading Marx, Kropotkin, and Luxemburg and from then on I've always had a left wing orientation.

Steve eventually went to university which was unusual for his family since at that time "*the idea of going to university was quite bourgeois – one didn't do that sort of thing*". I would like to suggest that this history becomes quite significant in Steve's subsequent orientation to social justice. This orientation, along with his Jewish cultural heritage, led Steve and his wife Beryl (as Gilah Leder writes in her chapter) toward Israel and the kibbutz movement in 1970.

> In my youth I was a socialist and read Marx, the anarchists, the social democrats, and all that stuff and was very concerned about equity. I saw going to a socialist community, kibbutz, as playing a part in that. It just made sense to live in a kibbutz which was as much socialist as Zionist. When I went to live in Israel, it was important to us to work on the land – and I lived on a kibbutz for 6 months.

Steve the Teacher

However, things didn't start off too well, and Steve was very soon "*surplus to requirements*" on the farm though he does admit this was probably because he was a pretty poor farmer. Fortuitously for all of us, his need to be usefully occupied, led to an opportunity to get involved in teaching. This lasted as long as he stayed in the kibbutz, but it made such an impact on him, that he decided teaching – not farming – was the occupation for him.

> We left the kibbutz in 1973 to begin teaching maths in London; I got a job at a very progressive private school called King Alfred's. For a year and a half I really enjoyed it. They called us by our first names, and we sat with the kids at lunchtime. That school was very privileged, and I was a bit overawed meeting Peter Sellers and Linda McCartney at

parents' evenings. I moved from there to teach maths in another school in Hendon, St Mary's. The intake there was skewed toward those with lower social class backgrounds. Within six months I must have proved my worth because they asked me to apply for the vacant head of maths job – which I got. There we wrote our own Mode 3 CSE course and set our own final exam.

It is very difficult describing to today's teachers that in the 1980s in the UK, it was possible that teachers not only designed the school curriculum, but also designed and marked the assessments including final examinations. These days (in the 2010s), teachers are no longer trusted by our politicians but require constant haranguing, criticising and monitoring by individuals who have barely set foot in a state school classroom. Yet in the 1980s it was possible to have the freedom to work within a school subject department in ways that fitted a commitment to social justice – something which Steve worked for in his early career.

For me then it was all about trying to do something to help ALL kids achieve in maths. I'd like to think we had an effect, but I don't know. I have no statistics. We had more kids going on to A Level though, so that must say something. I taught there for five years. I instigated department meetings to decide democratically on all things so that staff would buy in to the decisions we took. This involved a big change as it had been quite a traditional school. They had setting but I didn't succeed in fighting that – it was too big a step for the staff to take. We did try to make sure we gave the lower sets a credible maths course though. Concern about equity was something I attended to, insisting as head of maths that all staff taught all sets, and that we used SMILE as I thought that would play a useful role too. But at that time I had no theoretical tools to translate a concern for equity into actually understanding how we could be doing something about it. It is clear here that Steve takes a position in opposition to setting – also called "tracking" in the United States. This is that practice in the UK whereby children are placed into different mathematics groups according to some process of "assessment" though in practice means pupils become stratified within the school by social class (see Jorgensen et al. 2014). Steve's opposition to setting here is important, since setting is centrally implicit in the underachievement of disadvantaged pupils (Boaler 1997; Boaler et al. 2000). One just simply cannot claim to support equity and social justice and not be opposed to such discriminatory practices which lie at the very roots of inequity. Robyn Zevenbergen (now Jorgensen) has discussed and theorised this in (Zevenbergen 2003). That Steve used SMILE (Secondary Mathematics Individualised Learning Experience) is also significant. SMILE was initially developed as a series of practical activities for secondary school students by practising teachers in London the 1970s. It spread rapidly but provided a creative approach to the learning of mathematics through activity and exploration quite different from the traditional textbook approach.

After 3 years or so at this school, Steve began looking outside of the school for inspiration.

In 1979, I quickly got drawn into Alan Bishop's work, and he was one of the key influences of the time, looking into culture, values and so on. Alan comes from a very British liberal tradition, but I felt he shied away from critical issues and from even talking politics about Marxism and socialism. He focussed on how can we do good by kids, but without digging away at what caused the problem they faced. I had enormous respect for him, and still do, but I feel if we don't challenge it, then we won't change it; if we work on surface we don't change anything.

Between 1979 and 1987 Steve's career began shifting: 1979–1982 PhD at Kings; 1984 PGCE tutor at the London Institute of Education (IOE). However, he was rejected by the IOE for a permanent post in 1987 in a decision reminiscent of Decca records rejecting the Beatles in 1962 saying "*Groups with guitars are on the way out. The Beatles have no future in show business*" (http://en.wikipedia.org/wiki/The_Beatles'_Decca_audition). But thanks to Ros Scott-Hodgetts, in 1988 Steve applied for and got a post as Lecturer in Mathematics Education at London South Bank University, where he stayed until 2013 becoming Professor in 1998 and Emeritus in 2013. For those of us who have made somewhat similar transitions, those were heady days, where one became exhilarated at the intellectual challenge and the opportunities one was handed. Of course, the down side is it removes you from the everyday struggle; we have the privilege of intellectual engagement in the big ideas, but lack the engagement in the activity.

Whilst Steve appears to have always been progressive by inclination, his intellectual tools were still in a process of development.

> In my PhD social justice was a bit second-hand to be honest, however all the sort of systems of belief would influence what mathematics you taught and how you taught, so though it was not foregrounded, it was in the background. Working with Paul Hurst, him being a philosopher, I started looking at the philosophy of mathematics; he made me think there must be a connection between what you believed and how you taught. Then I met a guy called Paul Ernest and we started talking and it snowballed from there.
>
> At that time I had not encountered many people in maths ed. I had been to some of the Chelsea Seminars where the great and the good went, so it was a real eye opener, seeing some of these names I'd seen on the back of the books. People must do that now... look there's Steve Lerman, but I'm just Steve, I still shake my head and wonder why.
>
> I first went to a conference in 1984 in PME8 in Australia. In academia, I again addressed equity but still no real idea. Then working with Anna Tsatsaroni I came to understand from the Marxist sociologists of education how inequity was produced and therefore how it may be possible, within the constraints of schooling in capitalist society, to educate teachers to have some understanding of the production of inequity and have some effect. It feels now that there's a kind of joining up from my youth to how I think now.

So moving to South Bank was not merely an accident of history; as an institution it had a vision that was – and still is – consistent with a social justice orientation, something with which Steve already had an affinity:

> At South Bank it was easy to feel comfortable. The majority of students were from the local population, from low socio-economic backgrounds, from minorities, with people coming in as second chance learners. In 1987 I ran the MSc in Mathematics Education. Part of the approach was that the maths teaching had to be good, informed by research then, we pushed their maths into new areas, but also not accepting children opting out, that social class was a key reason for why some kids were opting out. Students came out inspired and committed to doing something about it. They are going out saying we can do something and make a difference
>
> I had the privilege in working with a really good head of department; an old leftie who took over in 2000. She got us spending days talking about values. We sent our students into inner city schools to teach disadvantaged kids and we see them still coming out disadvantaged. The question we were challenged with was: how can we bring our values into our teaching? We started courses on equality, citizenship and inclusion. We bought in experts on class, sexuality and so on to inspire teachers. When we eventually set up the EdD we actually made it a theme of our doctorate. I think we had a full cohort exactly because it attracted people working in areas of social inclusion.

So Steve had found his intellectual home, possibly explaining his stay of 25 years.

Steve on the Turn

Steve's progression to an effective sociologist of mathematics education, came through a number of opportunities that allowed him to get closer to others with similar orientations.

> It must have been 1996/7, when Joâo Filipé Matos from Portugal got funding for a research project, to work theoretically together to see what we might make sense of in how to change mathematics in schools. This bought together Candia Morgan, Magdalena dos Santos, Jeff Evans, Anna Tsatsaroni. I learned so much and began to look at Bernstein through Anna's pressurising!

One skill Steve has, from which I have gained considerably, is his ability to weave together diverse strands and forge a coherent argument with a clarity that puts the focus on the underlying ideas (rather than on some imagined intellectual supremacy of the author which we see in rather too many of our contemporaries). It is in this way I believe Steve lives out his commitment to social justice. Steve seems unconcerned with writing in order to appear clever, or to distance himself from the reader by obfuscation. Rather his writing seems concerned with giving people tools for thinking and tools for action. His writing is thus democratic, not autocratic, inviting people to draw their own conclusions by entering into his argument and potentially seeing things, if not differently, then more clearly. In that way he contributes to the community.

One of the key contributions Steve has made to our field and which has provided tools for thinking and action has been his "Social Turn" chapter (Lerman 2000) in Jo Boaler's book *Multiple Perspectives on Mathematics Teaching and Learning* (Boaler 2000). This chapter however does not address equity issues, or social justice, nor does it address social class as a key driver of mathematical achievement. Yet what it has done is influence people in mathematics education and clarified the intellectual legitimacy of looking beyond the individual as cognising subject. Yet to take this chapter on face value and conclude that Steve's spirit was unconcerned with social class is to miss the politics behind the polemic.

> Yes, I do think the social turn is an important perspective, that focusing on cognition without thinking of the whole child is fine for advantaged children, but that only perpetuates disadvantage. You kind of hope that the whole field will find that they have to think about the whole social setting of education, even when looking at using dynamic geometry to learn concepts. But they don't.

I am only too aware that many in the field do not confront endemic injustice. Yet whilst I might have a tendency to be somewhat confrontational, Steve's tendency is to work hard to convince by providing tools for thinking.

Steve and the Left

> I mourn the losing of the revolutionary battle. But Marxist thought is still there through Vygotsky and Leontiev, also through Bernstein and Bourdieu, We are talking again about Marxism but it's not enough. I wasn't a revolutionary Marxist. The kibbutz was egalitarian and I guess I was a small r revolutionary – being committed to egalitarianism, but the kibbutz didn't change the country.

One of the frustrations those of us on the left experience is that a lot of research in mathematics education is class blind and does not take into account children's backgrounds. Yet millions of children don't attend school, and of those who do, millions don't actually sit in the sort of classrooms that get all the research focus. In some ways then our mathematics education research is an elite minority sport. There are important exceptions to this through a significant strand in critical mathematics education – something which Steve acknowledges as a key influence on his development in the 1980s:

> The people influencing me at the time were Claudia Zaslavsky Marilyn Frankenstein, Arthur Powell, Gelsa Knijnik.

These influences were evident in some of Steve's most early work (see Lerman 1989a, 1989b, 1993). As a Marxist, for me the most exciting of Steve's papers has always been his own contribution to the book he co-edited with Marilyn Nickson – "*The Social Context of Mathematics Education*" (Lerman and Nickson 1992). The chapter "*Learning Mathematics as a Revolutionary Activity*" (Lerman 1992), must be one of the most tantalisingly political titles – vying with Robyn Zevenbergen's "*Constructivism as a Liberal Bourgeois Discourse*" (Zevenbergen 1996). Of course at that time, Steve was still an angry young man (Well, "two out of three ain't bad"). Sadly there is no Marx or Lenin in the references, so it can't be *that* revolutionary. Steve, however, makes the claim in this chapter that mathematics teaching through problem solving can be "emancipatory".

This would, however, require a major shift in how teachers perceive their role because, "*teachers maintain the passive acquiescence of the oppressed in society*" (Lerman 1992, p. 171) Challenging this with slogans is not enough but one has to "*reveal the fundamentally revolutionary nature of the alternative and become actively involved in it*" (op. cit.). But, given the place state education has in the maintenance of capitalism, assuming a revolutionary role has to be *subversive*. One has to watch one's back, because a desire to fundamentally change society will mean those with power will have some of theirs, if not all of it, taken away.

In this 1992 chapter, and elsewhere (Lerman 1983), Steve rejects the notion of a value-free mathematics and mathematics education – a position which went on to form the basis of his PhD (Lerman 1986). This remains a contested standpoint in our field, and the innocence of mathematics and mathematics education still pervades more deeply than we care to admit. What is perhaps more insidious than the widespread rejection of mathematics as a social construction, is the acceptance in our field of both the existence and the immutability of "ability". Even Steve, slipped into this by accepting our values will influence our decisions especially when "*we may have to consider what mathematics we teach to pupils of different abilities*" (Lerman 1992, p. 175). It is a standard position within the sociology of education (though I admit it might not have been so widespread in 1992) that mathematics education practices construct the ability level of pupils through the curriculum and forms of pedagogy, as well as the responses to pupils from home backgrounds other than the middle class culture favoured by schools (See the works of Basil Bernstein, Michael Apple and Pierre Bourdieu). So the acceptance of pupil ability as a structuring characteristic of the curriculum is already an acquiescence to the oppression of working class kids.

However, an important step Steve takes (1983, p. 176) is to highlight the way in which the introduction of problem-solving into the mathematics classroom, might well bring about no change at all, if the teacher is not aware of the need for what is the fundamentally different activity that is required. In addition, the problems to be solved in themselves need to be emancipatory, allowing teacher and pupil to own the mathematics and to see the way in which mathematics can be used to analyse, criticise and ultimately change the world at large (p. 197). This not only means we select "good" problems (now the fashion is to call this "task design") but that we make it explicit that the act of engaging in problem solving, asking questions examining data, etc., is itself emancipatory and therefore a challenge to the authoritarian nature of schools, classrooms, curriculum, pedagogy and assessment. This position is more fully taken up by Rico Gutstein and others in addressing the question: why do we engage in problem solving if it is not to change the world and more specifically to illuminate and change the direct oppression felt by pupils and their communities (Gutstein 2003, 2006; Gutstein and Peterson 2005)? One thing that is probably never asked in such classrooms is:why are we doing this?

Continuing with the theme of empowerment, in 1995 Steve gave a plenary lecture to the British Congress of Mathematics Education (BCME), titled "*Mathematics Teaching for Empowerment and Equity*". In this he looked at various levels of established practices, but he tackled one aspect of the debate over equity and mathematics: do we focus on the empowerment of disadvantaged students, or on the mathematics we teach them? By drawing on the work of Gelsa Knijnik with the landless peasant movement in Brazil, Steve's stance was to argue that pupils become empowered *through* mathematics.

> My concern is with equity through learning, not through a differentiated curriculum or posited on differentiated needs according to unchangeable notions of ability. I am arguing that it isn't adequate to work towards each child achieving the best of her/his potential because that is inspired by the expectation that those who have the ability, or those who have the appropriate class/gender/ethnic background, will succeed at a higher level than others, however much we may regret and even oppose such institutionalised disadvantage. (Lerman 1995)
>
> In the mathematics classroom I want to propose that our aim has to be the success of more pupils and to that end the focus has to be on equity, which I take to mean that all pupils have an equal chance to enjoy mathematics and succeed in standard assessment tasks too, if as teachers we can find appropriate ways to realise that. (Lerman 1995). Steve took this further in his contribution to a collection edited by the late Leone Burton where he pulls together various theoretical strands in order to provide a framework where psychology can "take account of political, social and economic issues" (Lerman 1999, p. 95).

Steve, PME, MES and Beyond

Steve has been a central figure in PME for many years, and served as its president between July 1995 and July 1998. Steve's first thought on being elected was "*my god I could ruin the whole thing*". As we now know, he didn't.

At successive PME conferences, I have experienced how rooms fill up when Steve's name is on the programme with many people believing he will have something valuable that speaks to them.

> Being in PME has been an inspiration in my academic and social life. A discipline and meeting good people. In terms of social justice. It is always nice to be on the fringe sometimes, and in terms of work around social justice and sociology, important to be a forum to meet people with MES starting. As president in 1995, the conferences in 1996 and 1997 were already set out but in 1998 my first that I could influence – in South Africa – we were able to interview Michael Apple; where else might we be able to do that? That was a big issue... Social issues were central.

Whilst being a central pillar of PME, Steve has also supported change in its orientation, taking a number of years.

> As president I encouraged setting up a discussion in 1999 to discuss a change in policy, but it was not the right time. But by 2005 it was the right time being in Australia – because where it is makes a huge difference. Maybe Brazil would have been right but some other paces – Japan, for example – it would not have been right.
>
> Why is it the case that social issues were not central to such an important organisation? Well sometimes there are other places people go to but on the negative, there is an inertia. There seems to still be a psychological focus. Those supervising new PhD students still have an influence. It is hard to understand because to us, it is so central to our thinking.

At the AGM of PME27 (in Hawaii), Steve seconded a motion that I proposed which called on the International Committee to initiate a discussion about the future of PME and to bring a report to a future AGM. This was because the strict focus on the "P" (psychology) no longer represented the work of many PME members. At the time I argued that

> ...it is clear that the discipline has shifted over the years so a tight focus in psychology is no longer appropriate. Certainly there is a problem over papers that are rejected (rightly as it happens) just because they are not psychological. The debate is then polarized over "psychology" or not. (Gates 2004)

Having Steve, as an ex-PME president, in clear support I believe gave the proposal greater credibility. This proposal was first discussed and passed in PME28 Bergen in 2004, and Steve proposed the ratification of the motion in PME29 Melbourne in 2005 (I was unable to attend, so I left him to it!). In his presentation to the AGM Steve highlighted the need to retain the success and experience of PME and the need for PME to formally recognize the broad and wide theories that members currently draw on in their research. But significantly also highlighted how the proposed constitutional change would "*ensure that PME foregrounds its willingness and openness to work on issues of equity and social justice, most of which draws from theories outside psychology*" (PME29 AGM Minutes).

So was Steve satisfied we has bought about a significant change in how PME operated and structured the discipline? No, not really.

> I have to say I have been disappointed in what little change there has been in PME. There were lots of us around lots of new people really pleased with the change in emphasis. But still social class is not there. I have a sense the UK community, and maybe Australia is more aware of social class than the international community. You know in the US you can't talk about class. There is more awareness now but I'm not sure it figures in academic papers.

In his contribution to the PME Handbook (Gutiérrez and Boero 2006), which was being written between PME 28 and PME 29 Steve was invited to look at socio-cultural research in PME (Lerman 2006) and was able to discuss the place of sociology and socio-cultural approaches within the history of PME. Equity and social justice took a back seat as this was explicitly discussed in another chapter (Gates 2006). However Steve does raise political issues within the argument for a greater role for socio-cultural theories, whist at the same time suggesting we needed to adopt a more diverse understand of "psychology" (Lerman 2006, p. 349). Whilst Steve is clear that his attention is largely on a social-cultural analysis at the level of classroom interactions, he also recognises the importance of taking a broader perspective. In addition (p. 357) he raises the importance of studies which identify the ways in which some invisible pedagogies create classroom interactions that disadvantage students from low socio-economic backgrounds. He raised a number of issues – learner identity, studies of the pedagogy of mathematics classrooms, mathematics teacher learning as well as "*what it means to be school mathematical*" (p. 363). This still positions Steve's analysis in this chapter within the school culture and community – something which is consistent with his affinity for Bernstein. However his final paragraph in that chapter raises a much bigger question:

> . . .who fails in school mathematics, and how do they fail, and why? In many countries around the world particular social groups are associated with failure in school mathematics, determined by social class, ethnicity, poverty, gender . . . (Lerman 2006, p. 363)

Here the notion of social class is explicit and political struggle inherent.

In the Mathematics Education and Society Conference in Portugal 2008 (MES5), Steve, along with Andri Marcou, presented a paper that aimed to present a model of the various approaches to studying equity in mathematics education with a view to mapping the field and enabling a conversation between different perspectives (Lerman and Marcou 2008). They conjectured that in their study (admittedly of only 11 projects reported in journals) there was a higher percentage of papers that did *not* draw explicitly on a theory, than those that did (Lerman and Marcou 2008, p. 8), conjecturing that this might have been in part due to "*the political pressure to focus on 'what works' in educational research in the USA*", as well as a preponderance of "design research" which acts as a pragmatic response to making things work in education. They wrote:

> We are convinced that if we have some good explanations for why a specific intervention or feature works it may be translatable into different circumstances by different teachers at different times, as it may be adaptable in its application to those circumstances. Some theories provide a rich explanatory tool; in research reported above, we have found

> Bernstein's theories to be just such a framework (see also Morgan et al. 2002). In other approaches explanations are perhaps harder to come by. (Lerman and Marcou 2008 p. 9)

In 2006–2007 Steve took part in a UK Economic and Social Science Research (ESRC) research seminar series on *Mathematical Relationships: Identities and Participation*, resulting in a book in which Steve had a chapter (Lerman 2009). Steve though, along with presumably most academics, sees the role of scholarship to present, re-present and contextualise theory and he continually does this with clarity achieving what Bernstein appeared to be uninterested in doing: making the ideas clear and practical (see in particular Lerman & Zevenbergen 2004). What this seems to miss for me, however, is a stage of understanding the basis behind class oppression as the ideological armoury of the economic system. Steve was the only author in that book to mention – and even cite – Marx's position that social being determines consciousness. Marx gets another (brief) mention in Steve's 2001 ESM paper (Lerman 2001) – in fact it is to the same quote.

> It is not the consciousness of men that determines their being but, on the contrary, their social being that determines their consciousness. (Marx 1859, p. 328/9)

The significance of this is not lost on Steve – as it is on many who adopt a socio-cultural perspective – that it is fundamental we challenge the idea of:

> . . . the individual as the autonomous builder of her or his own subjectivity. Consciousness was to be seen as the result of social relations; in particular, relations to the means of production. (Lerman 2001, p. 89)

The second part of that quote being where many seem to part company with the political implications of the standpoint; that within capitalism, the working class will always be exploited *de facto*. Steve did not go so far as examining the ways in which a class analysis needs to be more explicit though. Steve appears to take the road which presumes we focus on the mathematics classroom in order to improve the potential of working class kids as distinct from improving the potential of all. This is also the case in Steve's now classic PME 22 paper "*A moment in the zoom of a lens*" (Lerman 1998). Steve's use of the "zoom lens" metaphor has been very attractive to many and is widely cited (Google gives me over half a million hits!). Yet I think he might do better with a wide angle lens as his zooming out stops in the mathematics classroom and so fails to engage with the wider social and political implications of those discriminatory practices he critiques so well. The area into which Steve's published work does not thus appear to take him is into critical mathematics education, though I do personally know that ideologically this is where Steve is most comfortable. This raises a question I have had for decades: why so many of my close professional associates, with whom I share much (though not all) of a political allegiance, chose to remain apolitical in their academic work. You know who you are!

I want to take issue with one further sentence Steve presents us with:

> Pupils' identities in relation to mathematics are, of course, largely formed in classrooms... although ... there may well be interaction outside the classroom that play a part in their mathematical identifies. (Lerman 2009, p. 150–151)

This is I think a step too far in trying to claim too great a responsibility for ourselves. Anyone familiar with attempts to teach working class kids (as Steve has actually done in his youth) must be aware that the biggest challenge resulting in an antagonistic mathematical identity has been the identities forged in the family and playgrounds, which place the mathematics classroom a poor third at best. I would argue strongly that a focus strictly on the classroom as the place for improvement is myopic. Studies have shown that poverty has a much stronger influence on school achievement than instructional quality, leading to a policy imperative that if we want all pupils to do well "*minimizing social inequities must be a fundamental component of education policy*" (Georges 2009). We can go further:

> The type of neighbourhood in which a pupil lives is a more reliable predictor of a pupil's GCSE performance than any other information... The performance of pupils from any particular type of neighbourhood is also incrementally affected by the neighbourhoods from which the other pupils in the school they attend are drawn. (Webber and Butler 2007)

The problem with identity (and most of postmodernism actually) is that it again allows one to be apolitical – an identity is something you have, whereas a critical perspective focusses us on ideology as something deeply rooted within a class consciousness. If working-class kids are to improve, it needs to be at the *expense* of affluent kids, otherwise it is no improvement at all. This gaze onto identity politics, at the expense of real politics, seems at odds however with where Steve's heart seems to have been. Steve has had a commitment for nearly 25 years to working in London South Bank University, working with, and on behalf of, disadvantaged communities.

Finishing Off

So, given the often insurmountable problems we face in bringing about a more just mathematics education, what would Steve do differently?

> I think we need to engage with policy and policy makers much more explicitly. If you don't talk about values, then by omission you are talking about values.
>
> What by omission can teachers do... given where we are? Well, it's making the very best of a bad job. We have to find a say for kids from disadvantaged backgrounds to enjoy maths in order to get through and gain qualifications to make better choices, but also to understand where assumptions about ability come from and why they are where they are. Also to become angry and to direct that anger and do something about it.

Steve's writing might not always have foregrounded equity and social justice, but he has given us many of the tools needed to take that commitment further. Most importantly, he has given us the credibility to ask fundamental questions about why some groups of learners succeed at the expense of others. Finally he has made it clear, that those of us on the left are the nice guys!

Let me finish with Steve's reflection on two questions. First, the question of whether or not we make a difference?

> Does it make a difference? I fear not as much as we think.
>
> I think we are battling against governments, Labour as well as Tory. If I were starting again, I would want to have more influence in schools by working on and challenging policy. Maths ed community does not have a public focus, we look inwards. Whilst some do that (Frankenstein, Knijnik) most don't. We bemoan what policy does. Some of those that do generally end up supporting the status quo, producing material and ideas that support policy. Do materials support current policy or challenge it? There is no research that says good calculator usage is any way bad for kids. So why do we not challenge that? Similar for setting. For example, this setting thing is ruining it – we know who slips down and gets ignored. We know in practice its "keep them behaved, but don't give them interesting stuff". But I can't help but feel our student teachers come out inspired and determined to do something. But, you can't give up.

Finally, I return to the question I posed at the beginning of this chapter, how Steve himself feels his work has befitted the most disadvantaged in society.

> I like to think that in responding positively to things I and others have written about the social turn and a sociological perspective, which many people have done, has opened minds to another way of understanding who fails and why. But most of all it has helped me see things differently and in my teaching I always raise the issue, and again receive positive responses from students who are going to teach in inner London schools. If it has made any difference in practice I can't know of course. I would hope any changed attitudes and practice would lead to a widening of who succeeds, I doubt it would be at the expense of the advantaged. Bernstein shows how the middle classes always succeed, whatever the pedagogic form.

Speaking personally, I have been very highly influenced by Steve's thinking, and it helps that Steve is a thoroughly decent bloke.

Acknowledgement I want to publically thank Peter Winbourne for his very useful comments and suggestions on earlier drafts of this chapter.

References

Boaler, J. (1997). Setting, social class and the survival of the quickest. *British Educational Research Journal, 23*(5), 575–595.

Boaler, J. (Ed.). (2000). *Multiple perspectives on mathematics teaching & learning*. Westport: Ablex Publishing.

Boaler, J., Wiliam, D., & Brown, M. (2000). Students' experiences of ability grouping – Disaffection, polarisation and the construction of failure. *British Educational Research Journal, 26* (5), 631–648.

Gates, P. (2004). *The future of PME* (A Report on the Issues by the International Committee of PME). PME.

Gates, P. (2006). The place of equity and social justice in the history of PME. In A. Gutierrez & P. Boero (Eds.), *Handbook of research on the psychology of mathematics education. Past, present and future* (pp. 367–402). Rotterdam: Sense Publishers. ISBN 90-77874-19-4.

Gates, P. (2014). Equity and access in mathematics education. In S. Lerman (Ed.), *Encyclopaedia of mathematics education*. Springer.

Gates, P., & Jorgensen, R. (2009). Foregrounding social justice in mathematics teacher education. *Journal of Mathematics Teacher Education, 12*(3), 161–170. doi:10.1007/s10857-009-9105-4.

Georges, A. (2009). Relation of instruction and poverty to mathematics achievement gains during Kindergarten. *Teachers College Record, 111*(9), 2148–2178.

Gutiérrez, A., & Boero, P. (Eds.). (2006). *Handbook of research on the psychology of mathematics education* (pp. 347–366). Rotterdam: Sense Publishers.

Gutstein, E. (2003). Teaching and learning for social justice in an urban Latino School. *Journal for Research in Mathematics Education, 34*(1), 37–73.

Gutstein, E. (2006). *Reading and writing the world with mathematics: Toward a pedagogy for social justice*. New York: Routledge.

Gutstein, E., & Peterson, B. (Eds.). (2005). *Rethinking mathematics: Teaching social justice by the numbers*. Milwaukee: Rethinking School Ltd.

Jorgensen, R., Gates, P., & Roper, V. (2014). Structural exclusion through school mathematics: Using Bourdieu to understand mathematics a social practice. *Educational Studies in Mathematics*. doi:10.1007/s10649-013-9468-4.

Lerman, S. (1983). Problem solving or knowledge centred. The influence of philosophy on mathematics teaching. *International Journal of Mathematical Education in Science and Technology, 14*(1), 59–66.

Lerman, S. (1986). *Alternative perspectives on the nature of mathematics and the possible influence on the teaching of mathematics*. Unpublished PhD thesis, Kings College, University of London.

Lerman, S. (1989a). A social view of mathematics – Implications for mathematics education. In C. Keitel (Ed.), *Mathematics education and society* (pp. 42–44). Paris: UNESCO.

Lerman, S. (1989b). Investigations, where to now? In P. Ernest (Ed.), *Mathematics teaching: The state of the art* (pp. 73–80). Lewes: Falmer Press.

Lerman, S. (1992). Learning mathematics as a revolutionary activity. In S. Lerman & M. Nickson (Eds.), *The social context of mathematics education: Theory and practice* (pp. 170–179). London: Southbank Press.

Lerman, S. (1993). Critical/humanistic mathematics education: A role for history? In M. Selinger (Ed.), *Teaching mathematics* (pp. 97–102). London: Hodder & Stoughton.

Lerman, S. (1995). *Mathematics teaching for empowerment and equity: An achievable goal?* Invited Plenary Lecture, BCME.

Lerman, S. (1998). A moment in the zoom of a lens: Towards a discursive psychology of mathematics teaching and learning. In A. Olivier & K. Newstead (Eds.), *Proceedings of the twenty-second conference of the international group for the Psychology of Mathematics Education* (Vol. 1, pp. 66–81). PME Program Committee, Stellenbosch.

Lerman, S. (1999). Culturally situated knowledge and the problem of transfer in the learning of mathematics. In L. Burton (Ed.), *From hierarchies to networks in mathematics education* (pp. 93–107). London: Falmer.

Lerman, S. (2000). The social turn in mathematics education research. In J. Boaler (Ed.), *Multiple perspectives on mathematics teaching and learning* (pp. 19–44). Westport: Ablex Publishing.

Lerman, S. (2001). Cultural discursive psychology: A sociocultural approach to studying the teaching and learning of mathematics. *Educational Studies in Mathematics, 46*, 87–113.

Lerman, S. (2006). Socio-cultural research in PME: Perspectives, trends and results. In A. Gutiérrez & P. Boero (Eds.), *Handbook of research on the psychology of mathematics education* (pp. 347–366). Rotterdam: Sense Publishers.

Lerman, S. (2009). Pedagogy, discourse and identity. In L. Black, H. Mendick, & Y. Soloman (Eds.), *Mathematical relationships in education* (pp. 1247–1255). London: Routledge/Falmer.

Lerman, S., & Marcou, A. (2008). Are all studies on equity in school mathematics equal? In J. F. Matos, P. Valero, & K. Yasukawa (Eds.), *Proceedings of the fifth international mathematics education and society conference*. Lisbon: Centro de Investigação em Educação, Universidade de Lisboa – Department of Education, Learning and Philosophy, Aalborg University. http://mes5.learning.aau.dk/Papers/Lerman_Marcou.pdf

Lerman, S., & Zevenbergen, R. (2004). The socio-political context of the mathematics classroom: Using Bernstein's theoretical framework to understand classroom communications. In P. Valero & R. Zevenbergen (Eds.), *Researching the socio-political dimensions of mathematics education: Issues of power in theory and methodology* (pp. 27–42). Dordrecht: Kluwer.

Marx, K. (1859). A preface to contribution to the critique of political economy. In *Marx and Engels selected works* (pp. 361–365). London: Lawrence and Wishart.

Morgan, C., Tsatsaroni, A., & Lerman, S. (2002). Mathematics teachers' positions and practices in discourses of assessment. *British Journal of Sociology of Education, 23*(3), 445–461.

Webber, R., & Butler, T. (2007). Classifying pupils by where they live: How well does this predict variations in their GCSE results? *Urban Studies, 44*(7), 1229–1253.

Zevenbergen, R. (1996). Constructivism as a liberal Bourgeois discourse. *Educational Studies in Mathematics, 31*(1/2), 95–113.

Zevenbergen, R. (2003). Ability grouping in mathematics classrooms: A Bourdieuian analysis. *For the Learning of Mathematics, 23*(3), 5–10.

Chapter 5
Tracing the Advances in the Field of Mathematics Education

Charalampos Sakonidis

Introduction: Setting the Scene

The 1980s was a particularly important period for the coming of age of the field of mathematics education, which had just entered its second decade of developing as an autonomous scientific community. The mainly psychological and theoretical perspectives and methodologies dominating the previous years started being considered critically and the social, cultural and political issues began to be seen as significant for understanding pupils' success or failure in learning mathematics within a society and across societies. As a result, diversified theoretical models coming from a range of disciplines such as philosophy, sociology, anthropology and linguistics crept into the young discipline of mathematics education and were exploited by the research community in its attempt to formulate more effective analytical tools to make sense and intervene in the ways mathematics is learnt. Thus, sociocultural approaches little by little entered the field, fighting to become accepted and respected.

The 1980s was the time Steve Lerman first appeared in the community, studying originally at Chelsea College, then King's College London for his doctoral thesis (Lerman 1986), where one can trace the beginning of his interest in epistemological issues of mathematics education. In order to situate the course of his contribution to the field, I discuss some contextual determinants of its future direction at a national and an international level as well as of Steve's main contributions. In particular, I begin with a discussion of some influential state initiatives for improving teaching and learning mathematics in schools and research projects originated by these initiatives. Then, I present research activities which have had an impact on the field's developments and a commentary on scientific activities that signalled new directions in thinking about research and practice in mathematics teaching and

C. Sakonidis (✉)
Department of Primary Education, Democritus University of Thrace, Thrace, Greece
e-mail: xsakonid@eled.duth.gr

P. Gates, R. Jorgensen (Zevenbergen) (eds.), *Shifts in the Field of Mathematics Education*, DOI 10.1007/978-981-287-179-4_5

learning. Unavoidably, the above are mostly related to Anglo-Saxon countries, as relevant activities outside these countries made their appearance only from the 1990s, when socio-cultural approaches began to fight their way into the community.

Throughout the 1980s reports on low achievement in mathematics and concerns expressed publicly for the standards of the mathematics education offered in schools in the USA and in Europe led to the drawing up of agendas for action and to the setting up of committees to deal with these issues, thus establishing the need for change in the way mathematics is taught. The report "Curriculum and Evaluation Standards for School Mathematics" published by NCTM (1989) in the USA and the setting up of the Cockcroft Inquiry into the teaching of mathematics in UK schools (DES 1982) are two typical and very influential international examples of the actions undertaken to improve mathematics education standards. The former provided new objectives for students' learning, e.g., valuing mathematics, becoming confident in doing mathematics, reasoning and communicating mathematically and cooperating. The Cockcroft Report, on the other hand, recommended an emphasis on the uses of mathematics in everyday contexts, a broad description of a common core curriculum to prepare students for employment, a recognition of the need for curriculum differentiation, and a variety of teaching methods to include problem-solving, investigation, discussion and practical work, as well as exposition and practice. Both reports had considerable influence on mathematics teaching and assessment at all levels, but more importantly, fuelled a stream of research projects broadly following constructivism principles, the then emerging 'grand theory' for mathematics education (diSessa and Cobb 2004). Most of these projects, in fact, fit into one of the three constructivism traditions reported by Confrey and Kazak (2006), namely 'misconceptions, critical barriers and epistemological obstacles' (the other two being 'problem solving' and 'theories of cognitive development').

The relevant studies in this period carried out at Chelsea College (which merged with King's College in 1985) were among the most influential. The Concepts in Secondary Mathematics and Science (CSMS) project in particular investigated secondary students' mathematical and scientific reasoning aspiring to identify aspects of their thinking, rather than just measure achievement. The results made a very significant empirical and theoretical contribution to the documentation of children's understanding and misconceptions in school mathematics (Hart 1981) and subsequent projects sought to understand better the relationship between what was taught and what was learned (Johnson 1989). Subsequent national initiatives directed at improving mathematics teaching and learning drew on the CSMS study (e.g., the National Curriculum and the National Numeracy Strategy and Secondary Strategy). Nevertheless, the results of the above studies were criticized for a number of weaknesses. For example, the view of mathematics projected by CSMS findings as "an ordered hierarchy, and of pupils' mathematical abilities as correspondingly hierarchical" (Noss et al. 1989, p. 111), thus assuming that there exists a hierarchy of mathematics understanding, was seen as poorly justified and provocative. The criticism questioned the appropriateness as well as the limitations of the adopted theoretical and empirical approaches, thus keeping alive the quest for perspectives that would provide better understandings of the highly complex processes of learning and teaching mathematics.

Despite the efforts reported above, subsequent studies indicated that students' mathematical learning and thinking improved overall very little and in some cases even deteriorated abated. Large numbers of pupils kept failing in and being poorly related to mathematics (e.g., Hodgen et al. 2009). What is more, failure in and negative attitudes to mathematics were shown to be more persistent in minority populations, including girls and socially disadvantaged children (e.g., Walkerdine 1998). Thus, the challenge for theoretical frameworks and research methodologies that would go beyond psychological considerations to examine mathematics learning, incorporating social, cultural and political factors was open to be pursued. This quest began to be mirrored in contributions to scientific meetings organized around the world as well as in publications responding to calls for syntheses of past research and the setting up of research directions advancing the field.

For example, research reports and plenary/invited speeches underlying such a quest appear in PME's conference programmes. To mention just few but indicative occasions: (PME-10, 1986, London, UK), Seymour Papert /Beyond the Cognitive: the Other Face of Mathematics, Christine Keitel/Cultural Perspectives and Presuppositions in Psychology of Mathematics Education; (PME-12, 1988, Veszprem, Hungary), Terezinha Nunes/Street Mathematics and School Mathematics; (PME-14, 1990, Oxtabec, Mexico), Valerie Walkerdine/ Difference, Cognition and Mathematics Education. More importantly, a special day devoted to Mathematics Education and Society was reserved in the programme of the sixth International Congress in Mathematical Education in Budapest in 1988. The set of proceedings published afterwards (Keitel et al. 1989) is the first international collection of research papers on social factors in mathematics education.

Other publications indicating increasing awareness of sociocultural issues in mathematics education appeared in the Anglo-Saxon world. For instance, the first Handbook of Research on Mathematics Teaching (Grows 1992) included chapters on ethnomathematics (Nunes), gender (Leder) and race, social class and language (Secada). It was followed later by the International Handbook of Mathematics Education (Bishop et al. 1996) which tried to cover work not necessarily published in English. Lerman (2000) adds to this list publications referring to Vygotsky's work for the first time in Mathematics Education, e.g., Crawford (1981, 1988) and Cobb (1989).

The above reflect the community's efforts right from the beginning to identify theoretical and empirical approaches to understanding and interpreting the processes of learning and teaching mathematics in their entire complexity. It also shows the gradual, sometimes reluctant, awareness of the need for these approaches to incorporate psychological as well as sociocultural elements. Steve Lerman entered in the field as a member of a very active research community (British), exactly at the time that the movement to new perspectives was taking shape, enhanced by matured social demands (e.g., education for all), technological advances, such as microcomputers, as well as significant developments in related disciplines, like psychology (Vygotsky), sociology (Bernstein, Bourdieu) and philosophy (Foucault, Derrida, Habermas, Heidegger) to mention just a few. His work responded to the research and practice concerns of the community alike in an imaginative and knowledgeable manner, providing some challenging and occasionally provocative ideas. In the next section, an attempt is made to highlight some of his most influential ideas, particular those that had a significant impact on research in mathematics education around the world.

Some Influential Contributions

Three sets of ideas stand out in Steve Lerman's contribution to the field of mathematics education: epistemologies of mathematics and mathematics education, the turn to a socio-cultural view of mathematics education; and, mathematics teacher education and professional development. Of course, Steve's scientific activity covers a much wider range of issues, the bulk of which, however, I would argue, constitute developments along these ideas. In the following, I discuss some fundamental aspects of these central sets of ideas.

Epistemologies of Mathematics and Mathematics Education

At the time Steve Lerman entered the field, mathematics education research drew mainly upon two disciplines, mathematics and psychology, especially cognitive psychology. Mathematics provided a framework for analysing the knowledge to be pursued at school and psychology a rationale for the way to be effectively accomplished by children and teachers. An interest in epistemologies of mathematics and in the dominant constructivist paradigm as well as their influence on learning and teaching mathematics emerging at the time preoccupied Steve's early work.

Seeking to examine the influence of epistemologies of mathematics on mathematics education, Lerman (1990) identified two contrasting views of mathematical knowledge reflected in research and practice in mathematics education, that is, absolutism at one extreme and fallibilism at the other. In the former, mathematics knowledge is seen as absolute, certain, abstract and value-free, a discovery of timeless truths; the latter accepts the uncertainty of the mathematical knowledge, which is understood as a process of conjectures, proofs and refutations:

> Fallibilism, a view which accepts the potential refutation of all theories, and counter-examples to all concepts, allows one to ask how does one know that this answer is better than that one, what might constitute a notion of 'better', might they not both be possible, as with Euclidean and non-Euclidean geometries, or arithmetics with or without the Continuum Hypothesis. (Lerman 1989, p. 217)

For fallibilism, mathematics is the outcome of social processes, relative to time and place, characterized by its activity, which includes engaging in interesting problems, conjecturing, testing, reflecting, evaluating and communicating.

Recognizing that theories of mathematics have an important influence on theories of mathematics education, Steve focused analytically on relevant issues in a later work (Sierpinska and Lerman 1996). In particular, the authors attempted to elaborate critically the origins and to make explicit the basic assumptions underlying epistemologies in mathematics and in mathematics education. With respect to the latter, the subjective-objective character of mathematical knowledge, the role of social and cultural contexts in cognition and relations between language and knowledge were scrutinized. Also, dominant mathematics learning theories were compared and relationships between epistemology and a theory of instruction were explored. The authors concluded that "epistemologies of mathematics could find their way to mathematics education only via genetic, social, cultural and historico-critical epistemologies. Moreover …epistemologies do not translate directly into theories of instruction and do not make recommendations for the practice of teaching" (p. 867).

Based on the above, Lerman (1990) argued for the need for a closer interaction between philosophy and psychology of mathematics education, neglected for long because of the "predominance of interest in psychological aspects of what is taught and how" (p. 53). Trying to justify the strong association between mathematics and psychology characterizing the field of mathematics education, he identified two reasons (Lerman 2000). The first is related to the high status of the two disciplines, which offers a legitimization to the research carried out in the latter. The second reason refers to the fact that mathematics and mathematics education are strongly linked with the construction and preservation of the dominant systems of reason in the Western world. Mathematics is seen "as a marker of general intellectual capacity", allowing "its gendered and Eurocentric character, creating through its discursive practices the reasoning logical norm" (Lerman 2000, p. 21). Valero (2004), much in agreement with Lerman, attributes to psychology a crucial role in the process of reducing the student to a cognitive subject via mathematics education and the relevant research interest in his/her mathematical thinking processes. She concludes that "the discourses of mathematics education have resonated with the discourses of mathematics and psychology … in the construction of a particular research discipline, with particular theories and methods, supporting the constitution of practices in the classroom that fulfil essential social functions, which help in sustaining a certain kind of social organisation" (p. 5).

Among the psychological perspectives exploited in mathematics education, constructivism has been especially influential. Concentrating on its two hypotheses, that knowledge is actively constructed by the cognizing subject and coming to know is an adaptive process organizing the experiential world, Lerman (1989) noted that the shift of attention to teaching for understanding that constructivism brought about provided no answer to the question of how to make this happen and know that has happened. He attributed this to the notion of 'understanding' remaining tied to certainty and absolutism:

> …the process of coming to understand a concept is one that takes place in the mind of an individual, and the final step of achieving that full understanding of a timeless, universal notion is a very private, almost mystical one. It is certainly beyond the power of any outsider, such as a teacher, to know that the process has taken place in full (p. 221).

He argued that accepting the hypothesis that coming to know is an adaptive process organizing one's experiential world locates objectivity in the social domain: concepts and their meaning are public and so too is understanding: "theories and concepts are rooted in practice, and obtain their meaning from use. They gain their objectivity in their public nature, in that theories written down become public property, subject to dispute, negotiation and adaptation. Their objectivity does not lie in their being the ultimate truths" (p. 223).

In a later work, arguing against the central position of constructivism that the individual is the source of meaning, Lerman (1993) juxtaposed the work in cultural psychology, situated cognition, classroom studies and so on, arguing:

> Knowledge isn't in the individual's mind, nor 'out there' in objects or symbols. Knowledge is as people use it, in its context, as it carries individuals along in It and as it constructs those Individuals. Knowledge is fully cultural and social. And so too is what constitutes human consciousness. Communication drives conceptualisation (p. 23).

Thus, Steve progressively moved to a position of rejecting constructivism in favour of a fully socio-cultural view of the mind, the individual, learning and knowledge offering a much richer view of teaching and learning.

Sociocultural Perspectives and the Social Turn

In the 1990s, the shortcomings of drawing predominately on mathematics and cognitive psychology to understand the complexity of mathematics learning and teaching attracted the interest of a notable number of researchers. Constructivism had fueled the query of what it means to know and teach mathematics, placing the learner as an agent in the world. However, its individualistic approach to meaning-making (the individual was seen as autonomously building his/her own subjectivity) started being challenged on the grounds of the little attention paid to interpersonal and social characteristics of the learning context.

Steve Lerman joined the discussion expressing concerns about the adequacy of constructivism but also about attempts to consider the social and individual meaning making approaches as complementary. He argued that the main difference between Piaget's and Vygostky's theories is not that they down-play either the social or the individual, but that the one identifies the cognizing individual and the other cultural and discursive practices as the source of meaning respectively. That is, they rely on different premises with respect to the meaning making process (Lerman 1996). Shortly afterwards, espousing views on the individual as constituted in and through the social world, he suggested that:

> ...we might want to talk of the individual as a fragmented self at the intersection of a unique collection of overlapping identities constituted in different practices, as lived out through class, race, ethnic, sexual, gendered, regional and other positions. Thus 'we cannot fully specify the psychological subject/agent as an object whose nature can be defined in isolation from a context'. (Harre and Gillett 1994, p. 26 quoted in Lerman 1998a, p. 41)

Gradually but systematically Lerman's work was adopting a sociocultural perspective, where cultural and discursive practices become the source of meaning. In an overview of studies in mathematics education acknowledging 'social' factors, he referred to the growing body of such research as the 'social turn' in the field (Lerman 2000), describing it as "the emergence into the mathematics education community of theories that see meaning, thinking, and reasoning as products of social activity" (p. 23). These theories, like, for example, cultural psychology, theories of cognition in practice and sociological theories, recontextualized within the field of mathematics education led to the production of new knowledge. This knowledge differed from the one developed in mainstream mathematics education in that coming to know mathematics was not seen as emerging "from and within the mind of decontextualized cognitive subjects", but as "constituted in the encounter between contextualized, historically grounded human beings and their activity in particular settings and spaces that are socially constructed" (Valero 2004, p. 6). Lerman (2000) attributed the receptivity of the community to social theories to political concerns related to the continuing exclusion of some children from learning mathematics and the emerging impact of developments in disciplines like psychology – particularly of Vygotsky's ideas – and sociology, anthropology, political science, cultural studies and others on mathematics education community.

For Lerman (2000), the greatest challenge for research in mathematics education offered by theories within the sociocultural perspective is "to develop accounts that bring together agency, individual trajectories ... and the cultural, historical and social origins of the ways people think, behave, reason and understand the world" (p. 36). He proposed the metaphor of a zoom lens for research whereby the focus of study is seen as:

> ...a moment in socio-cultural studies, as a particular focusing of a lens, as a gaze which is as much aware of what is not being looked at ... Draw back in the zoom and the researcher looks at education in a particular society, at whole schools or whole classrooms; zoom back in and one focuses on some children, or some interactions... Research must find a way to take account of the other elements which come into focus throughout the zoom, wherever one chooses to stop. (Lerman 1998b, p. 67)

The move of psychology in the last three decades on learning in social practices and the way in which physical and cultural tools mediate learning turned the focus in mathematics education on discursive practices and on the social factors as constitutive of learning. In this view, learning is about participating in practices and becoming a member of a community where individuals are seen as discursively constituted. Such an approach enables the link between the actions of individuals and groups in the classroom, history and culture placing it within sociocultural tradition. As a person joins in a practice, the regulating effects of the practice begin, positioning the person in that practice. Thus, in searching for evidence of mathematical understanding, the focus moves on the students' developing identities as "speakers and actors of mathematics in school classrooms" (Lerman 2001a, p. 98). Hence, mathematical activities, texts, experiences, social relationships, positions and voices as well as histories and functions of mathematical artefacts acquire particular importance, being constituent elements of the students' identities.

The 'social turn' (Lerman 2000) in mathematics education brought to the fore new understandings and new concerns that gradually de-emphasise cognitive psychology as the only interpretative framework, favouring instead sociocultural theories. Our understanding of children's learning moved from a focus on the cognizing subject to being seen as a product of social activity, where not only the cognition of the subject but also his/her relations with other individuals and their shared discourses matter. According to Lerman, this shift was also the result of growing political concerns about the ways mathematics education is related to reproduction of inequalities through the structures of school, highlighting the critical role of mathematics education in society.

Mathematics Teachers' Education and Professional Development

A substantial part of Steve's work from the end of 1990s onwards is related to research on mathematics teachers' education, an emerging area of interest at the time and a pivotal concern of the community since then. In a review of the relevant studies Lerman (2001b) raised questions about the mainly implicit, not necessarily unproblematic assumptions of these studies with regard to teachers' learning. To substantiate his criticism, Steve focused on some central issues of the relevant research: teachers' beliefs and practices, reflective practice and teachers' knowledge.

In particular, he argued against the then identified trend to map teachers' beliefs to their practices as separated and stable entities. He suggested that they were both influenced by contexts; questioning the prevailing view that reflection promotes teachers' autonomous learning, critical judgement and freedom of bias. This prior view implicitly assumed a direction of development, ignoring the relationships of power which are present. Given that reflective practice takes place in social contexts Steve doubted the legitimacy of treating teachers' knowledge as the result of cognitive conflict situations, thus, implicitly extending learning through adaptation into adult learning. Summing up, he advocated that viewing teachers' learning as their growing awareness rests on a highly individualistic theory of learning which assumes that what is to be learnt is in one's head.

Based on his critique, Lerman (2001b) suggested an alternative route to mathematics teachers' learning, reflecting the emerging interest in social and cultural aspects of leaning in mathematics education at the time. According to this, learning is understood not as a cognitive organization and reorganization but as emerging identities through participation in various communities. This participation produces and not simply reflects beliefs, practices, purposes and goals, constituting the social settings which shape teachers' learning. Lerman indicated three theoretical perspectives as suitable for studying mathematics teachers' learning in socio-cultural terms: Lave's work (1988), activity theory and postmodernism.

In Lave's approach, the teacher-learner is seen as apprentice, who learns at the same time as becoming part of the social practice (e.g., teaching), thus developing identities. Wenger's (1998) description of identity as "a way of talking about how learning changes who we are" fits the dynamic character attributed to the notion of identity here as "something that is constantly negotiated through the interplay of one's lived experience in the world and how we and others discursively interpret that experience" (Goos 2013, p. 522). Within an activity theory perspective, human activity is seen as a system: a group of people who share a common object and motive over time, as well as the tools they use to act on that object and realize that motive; activity systems are constrained by division of labour and by rules. Tensions and contradictions within the system and between systems constitute sources of learning (Engeström 2001). Here teachers' learning is seen in relation to the social practices in which power and knowledge are situated. Finally, postmodern theory argues that teachers hold multiple, overlapping subjectivities. Each time, certain aspects of these subjectivities are called upon and are expressed by identities of powerfulness or powerlessness. Teachers develop new subjectivities as a result of their social and professional activities. Learning to teach here is viewed as becoming able to deconstruct practice and re-inscribe it into a language recognizing difference and enabling students' voices.

Lerman (2001b) argued that to study teachers' education we need theories that address the complexity of teaching as a social practice, recognize that research settings are themselves learning sites and to grant agency to teachers. Postmodern pedagogy, he advocates, appears to satisfy these requirements, encouraging the expression of difference, promoting methods of critique and encouraging learning and teaching theorizing. In a recent article, more than 10 years ahead and with a significant body of research on mathematics teachers' education within the socio-cultural perspective, he appears more careful, arguing for the need to explicate the focus of our theoretical lenses, warning, however, that expectations for coherence across research approaches and impact on learning and teaching practices should be kept low, because of varied social and cultural traditions as well as interpretations of educational research (Lerman 2013).

Sociocultural Research in Greece

Research activity in mathematics education in Greece was initiated by the end of 1980s. Among the first members of the community were researchers who had completed their postgraduate studies abroad and were already active in the international research scene. This group of people started to gradually build cooperation with teachers and researchers from Greece and abroad, creating the conditions for the establishment of today's vibrant community, with notable contributions in international meetings.

The timing of the emergence of Greek research in mathematics education in the international scene, the emphasis on developing partnerships with the international community, and the rigidity of the country's mathematics education system urging for reforms explain the choices shaping the relevant national research activity at the time. In particular, until early 2000s, although the dominant paradigm is constructivism, socio-cultural considerations and concerns enter the research agenda, mirroring a resonance to the 'social turn' emerging in the world community. Lerman's but also others' work undoubtedly contributed to this direction, as evidenced by his collaboration with Greek researchers (e.g., Lerman et al. 2002), the contributions to national meetings, the postgraduate Greek students and his frequently referenced publications.

In an effort to synthesize the Greek research activity in mathematics education of nearly 20 years at the 33rd IGPME in Thessaloniki, the broadening of the research perspective during the 1990s allowed viewing mathematics also as a sociocultural process was notified (Kynigos et al. 2009). Most of the studies reported under this research paradigm fall in the three main areas identified as crucial for Lerman's contribution to the field (see Section "Some Influential Contributions"). In particular, some of the studies examined teachers' instructional practices from an epistemological point of view (e.g., Tzekaki et al. 2002), identifying aspects of the way teachers manage the epistemological features of mathematics, which distort the mathematical meaning emerging in the classroom; other studies employed a variety of theoretical lenses, such as Vygotskyian or situated learning theories, shedding light on the decisive role of social and cultural practices and tools in the establishment of shared mathematical meanings (e.g., Kynigos and Theodosopoulou 2001). Finally, some studies focused on teachers' professional development, highlighting aspects of the crucial role of the interaction between teachers and researchers for this development (e.g., Sakonidis et al. 2007). In the following, I discuss three representative examples, one from the first and the other two from the third group of the above studies, aiming to exemplify the research developed along central dimensions of Steve's contribution to the field in the Greek context.

The first study reports on an attempt to examine the epistemological status of the mathematical knowledge interactively constituted in the classroom (Kaldrimidou et al. 2008). The basic position adopted in the study is that school and scientific mathematical knowledge differ (Sierpinska and Lerman 1996) and that the epistemological status of the former cannot be deduced only from the latter. It needs to be studied also in relation to the social contexts of the teaching and learning processes. To this direction, the authors focus on the classroom phenomena that determine the nature of the meaning emerging in the classroom and characterized as "mathematics". Three theoretical constructs were employed to investigate this nature: the idea of sociomathematical norms (Yackel and Cobb 1996), the notion of the epistemological triangle (Steinbring 2006) and the analysis of the management of the epistemological features of mathematics (Kaldrimidou et al. 2000). These constructs were used to analyse two mathematics lessons, provided by two secondary school teachers, in an attempt to examine the different features of the mathematical knowledge shaped in the classroom that each one of these constructs allows to identify. The results show that each of these three perspectives allows access to specific features of this knowledge, which do not coincide. Moreover, when considered simultaneously, the three perspectives offer a rather informed view of the status of the knowledge at hand. The authors concluded that the comparative and sometimes complementary use of different theoretical tools enables the sharpening of the analysis related to the mathematical status of this knowledge.

The second study is on professional learning developed in the context of longitudinal collaboration between three secondary mathematics teachers and two academic researchers on mathematics teaching practice (Potari et al. 2010). Conceptualising teaching as learning in practice and teacher change as a process of shifting participation in a community of inquiry (Jaworski 2006), the study focused on the reflective activity developed by its members as well as on the tensions and the conflicts that emerged in the shaping of an inquiry identified by them. The analysis of the data, mainly transcribed meetings and electronic communication, shows that during the years of working together, the collaboration between the teachers and the researchers changed in content (shifting from focusing on local practices to tackling broader issues of mathematics teaching) and in form (being transformed from a student–supervisor relationship based on a common interest in systematically inquiring into mathematics teaching to one of co-enquirers of its development). These changes can be attributed to the participants' commitment to explore each other's perspectives in relation to teaching mathematics, with the ultimate goal of arriving at a deeper understanding of it. This brought forward issues of 'belonging' and 'becoming', that is, of identity, and thus of identification with existing meanings and negotiation of new ones. As a consequence, tensions and disputes emerged, particularly at the beginning, usually related to the interplay between local mathematics teaching practice and its global context. However, at the same time, coalitions and alignments were enabled, forced by the requirement for some form of consensus in order for the whole process to be socially effective.

The third study concerns again professional learning but in a different site and through a different process compared to the above (Potari 2013). Specifically, the focus here was on the interactions between the research and the teaching activity of a group of prospective and practicing secondary mathematics teachers developed in the context of a Master's course in mathematics education. Taking the view that teachers' professional development is predominately a continuous process of critically aligning with the norms of the teaching community of practice, the author adopted a- third generation – Activity Theory perspective (Engeström and Sannino 2010) to identify links made by the participating teachers between the above two activity systems. The teachers were assigned a number of research-like and teaching tasks during one of the Masters modules aiming to initiate them to research in mathematics education as a tool for inquiring into teaching. The analysis of a group of five teachers' actions connected to the design and evaluation of a teaching intervention showed contradictions related to the rules and the division of labor in power in each of the two communities; also hesitant convergences emerging as the course was progressing with respect to a shared object of the two activity systems, that is, understanding students' thinking and exploiting research to inform teaching. To this direction, the role of the other teachers and especially of the teacher educator was crucial in promoting critical alignment with research practices, challenging reflective thinking on the interactions between researching and teaching and facilitating these interactions through the provision of appropriate tools.

The above studies are indicative of the research activity developed by members of the Greek research community within the sociocultural paradigm and along some of the central dimensions of Steve Lerman's contribution to the field. Of course, one should not neglect to mention work related to sociopolitical issues, which sprung amid the 'social turn', like, for example, identity and its political orientations (Chronaki 2013), or to cultural dimensions of mathematics education, e.g., language and culture interconnections (Stathopoulou and Kalabasis 2006). This body of research contributes to furthering the exploration of theoretical constructs and the accumulation of empirical evidence that advance our understanding of the dynamics of sociocultural readings of the "what" and "how" of mathematics education.

Concluding Remarks

Over the last three decades, the field of mathematics education addressed a number of important questions concerning research and practice in learning and teaching mathematics. As Jaworski (2006) points out:

> ...it has become common to think of mathematics in fallibilistic terms ... to consider learning as a constructive process ... to situate knowledge and learning relative to communities of practice ...and to debate the commensurability of constructivist and sociocultural learning theories ... Looking back ... we might refer to 'big theories' such as constructivism and sociocultural theories, that have been highly influential in addressing mathematical knowledge and the learning of mathematics. The mathematics education discipline has become mature in such theoretical considerations... (p. 188).

It is evident that Steve Lerman's work has been highly influential for this maturity to come about. He challenged individualistic, psychological theorizations to account for learning and teaching in mathematics classrooms, suggesting that these are social activities, situated within nested levels of socialization (e.g., schools, traditions) and influenced by a multitude of social factors (e.g., culture, race and gender). Within this perspective, he promoted a participatory model to account for coming to know and teach mathematics. This model focuses on the use of discourse and its contents (e.g., norms, values) as crucial mediating tools in order to interpret the mathematical learner in context.

Projecting in the future, Lerman (2006a, b) identifies three emerging key areas within the sociocultural paradigm which will develop further: learning as identity formation, activity theory and ethnography. In particular, learning and a sense of identity are the same phenomenon and the development of a school mathematics identity adds another layer to students' multiple identities, some of which are more important for them. Sociocultural perspectives appear to offer valuable frameworks to examining how the various modalities of teaching mathematics shape school mathematics identities. Activity theory on the other hand, places emphasis on the notion of human activity, incorporating cultural artefacts, social settings and the acting persons' goals and motives into a whole and on meaning as mediating the world for every individual through tools and signs, as well as through community, rules of activity and division of labour within the activity system. Finally, the interaction between practical and formal mathematical knowledge has attracted much research attention in the last decades. Relevant studies indicate that mathematics classrooms, like workplaces, are learning-in-practice sites characterised by tacit as well as explicit features and forms of participation and identity, thus needing ethnographic engagement to make sense of the emergent learning.

It is evident that the enormous complexity of mathematical learning and teaching processes leaves open the challenge of seeking to adequately understand them. Steve Lerman's work has resulted in a growing interest for exploring the social aspects of these processes, with the use of theoretical and methodological frameworks from a range of disciplines, such as sociology, anthropology, political science and cultural studies, contributing substantially to the direction of unfolding their complexity.

References

Bishop, A. J., Clements, K., Keitel, C., Kilpatrick, J., & Laborde, C. (1996). *International handbook of mathematics education*. Dordrecht: Kluwer Academic.

Chronaki, A. (2013). Identity work as a political space for change: The case of mathematics teaching through technology use. In M. Berger, K. Brodie, V. Frith, & K. le Roux (Eds.), *Proceedings of the 7th international mathematics education and society conference* (Vol. 1, pp. 1–18). Cape Town: MES.

Cobb, P. (1989). Experiential, cognitive, and anthropological perspectives in mathematics education. *For the Learning of Mathematics, 9*(2), 32–43.

Confrey, J., & Kazak, S. (2006). A thirty-year reflection on constructivism in mathematics education in PME. In A. Gitierrez & P. Boero (Eds.), *Handbook of research on the psychology of mathematics education* (pp. 305–345). Rotterdam/Taipei: Sense Publishers.

Crawford, K. (1981). Review of Wertsch. *Educational Studies in Mathematics, 16*(4), 431–433.

Crawford, K. (1988). New contexts for learning in mathematics. In A. Borbas (Ed.), *Proceedings of the 12th annual meeting of the international group for the psychology of mathematics education* (Vol. 1, pp. 239–246). Veszprem: PME.

DES. (1982). *Mathematics counts* (Cockcroft Report). London: HMSO.

diSessa, A., & Cobb, P. (2004). Ontological innovation and the role of theory in design experiments. *The Journal of the Learning Sciences, 13*(1), 77–103.

Engeström, Y. (2001). Expansive learning at work: Toward an activity theoretical reconceptualization. *Journal of Education and Work, 14*, 133–156.

Engeström, Y., & Sannino, A. (2010). Studies of expansive learning: Foundations, findings and future challenges. *Educational Research Review, 5*, 1–24.

Goos, M. (2013). Sociocultural perspectives in research on and with mathematics teachers: A zone theory approach. *ZDM – The International Journal on Mathematics Education, 45*(4), 521–533.

Grows, D. A. (Ed.). (1992). *Handbook of research on mathematics teaching and learning.* New York: McMillan Publishing Company.

Harre, R., & Gillett, G. (1994). *The discursive mind.* London: Sage.

Hart, K. M. (Ed.). (1981). *Children's understanding of mathematics* (pp. 11–16). London: John Murray.

Hodgen, J., Küchemann, D., Brown, M., & Coe, R. (2009). Children's understandings of algebra 30 years on. *Research in Mathematics Education, 11*(2), 193–194.

Jaworski, B. (2006). Theory and practice in mathematics teaching development: Critical inquiry as a mode of learning in teaching. *Journal of Mathematics Teacher Education, 9*(2), 187–211.

Johnson, D. C. (Ed.). (1989). *Children's mathematical frameworks 8-13: A study of classroom teaching*. Windsor: NFER-Nelson.

Kaldrimidou, M., Sakonidis, H., & Tzekaki, M. (2000). Epistemological features in the mathematics classroom: Algebra and geometry. In T. Nakahara & M. Koyama (Eds.), *Proceedings of the 24th conference for the psychology of mathematics education* (Vol. 3, pp. 111–118). Hiroshima: University of Hiroshima/PME.

Kaldrimidou, M., Sakonidis, H., & Tzekaki, M. (2008). Comparative readings of the nature of the mathematical knowledge under construction in the classroom. *ZDM – The International Journal on Mathematics Education, 40*, 235–245.

Keitel, C., Damerow, P., & Bishop, A. J. (Eds.). (1989). *Maths, education and society* (Document 35). Paris: Division of Science, Technical and Environment Education, UNESCO.

Kynigos, C., & Theodosopoulou, V. (2001). Synthesizing personal, interactionist and social norms perspectives to analyze student communication in a computer – based mathematical activity in the classroom. *Journal of Classroom Interaction, 36*(2), 63–73.

Kynigos, C., Philippou, G., Potari, D., & Sakonidis, H. (2009). Research in mathematics education in Greece and Cyprus. In M. Tzekaki, M. Kaldrimidou, & H. Sakonidis (Eds.), *Proceedings of the 33rd international conference for the psychology of mathematics education* (Vol. 1, pp. 303–322). Thessaloniki: PME.

Lave, J. (1988). *Cognition in practice: Mind, mathematics, and culture in everyday life.* Cambridge, UK: Cambridge University Press.

Lerman, S. (1986). *Alternative views of the nature of mathematics and their possible influence on the teaching of mathematics.* Unpublished PhD dissertation, King's College London.

Lerman, S. (1989). Constructivism, mathematics and mathematics education. *Educational Studies in Mathematics, 20*(2), 211–223.

Lerman, S. (1990). Alternative perspectives of the nature of mathematics and their influence on the teaching of mathematics. *British Educational Research Journal, 16*(1), 53–61.

Lerman, S. (1993). Can we talk about constructivism? In *BSRLM proceedings of the day conference* (Vol. 13, No. 3, pp. 20–23). Manchester: University of Manchester.

Lerman, S. (1996). Intersubjectivity in mathematics learning: A challenge to the radical constructivist paradigm? *Journal for Research in Mathematics Education, 27*(2), 133–150.

Lerman, S. (1998a). The intension/intention of teaching mathematics. In C. Kanes (Ed.), *Proceedings of mathematics education Research Group of Australasia* (Vol. 1, pp. 29–44). Gold Coast: Griffith University/Mathematics Education Research Group of Australasia.

Lerman, S. (1998b). A moment in the zoom of a lens: Towards a discursive psychology of mathematics teaching and learning. In A. Olivier & K. Newstead (Eds.), *Proceedings of the twenty-second annual meeting of the international group for the psychology of mathematics education* (Vol. 1, pp. 66–81). Stellenbosch: Faculty of Education, The University of Stellenbosch.

Lerman, S. (2000). The social turn in mathematics education research. In J. Boaler (Ed.), *Multiple perspectives on mathematics teaching and learning* (pp. 19–44). Westport: Ablex.

Lerman, S. (2001a). Cultural discursive psychology: A sociocultural approach to studying the teaching and learning of mathematics. *Educational Studies in Mathematics, 46*(1–3), 87–113.
Lerman, S. (2001b). A review of research perspectives on mathematics teacher education. In F. Lin & T. J. Cooney (Eds.), *Making sense of mathematics teacher education* (pp. 33–52). Dordrecht: Kluwer.
Lerman, S. (2006a). Cultural psychology, anthropology and sociology: The developing 'strong' social turn. In J. Maasz & W. Schloeglmann (Eds.), *New mathematics education research and practice* (pp. 171–188). Rotterdam: Sense Publishers.
Lerman, S. (2006b). Socio-cultural research in PME. In A. Gutiérrez & P. Boero (Eds.), *Handbook of research on the psychology of mathematics education: Past, present and future*. Rotterdam: Sense Publishers.
Lerman, S. (2013). Theories in practice: Mathematics teaching and mathematics teacher education. *ZDM – The International Journal on Mathematics Education, 45*(4), 623–631.
Lerman, S., Xu, G., & Tsatsaroni, A. (2002). Developing theories of mathematics education research: The ESM story. *Educational Studies in Mathematics, 51*, 23–40.
National Council of Teachers of Mathematics. (1989). *Curriculum and evaluation standards for school mathematics*. Reston: NCTM.
Noss, R., Goldstein, H., & Hoyles, C. (1989). Graded assessment and learning hierarchies in mathematics. *British Educational Research Journal, 15*(2), 109–120.
Potari, D. (2013). Promoting teachers' mathematical and pedagogical awareness. *Journal of Mathematics Teacher Education, 16*(2), 81–83.
Potari, D., Sakonidis, H., Chatzigoula, R., & Manaridis, A. (2010). Teachers' and Researchers' collaboration in analysing mathematics teaching: A context for reflection and development. *Journal of Mathematics Teacher Education, 13*(6), 473–485.
Sakonidis, H., Potari, D., Manaridis, A., & Chatzigoula, A. (2007). The collaboration between teachers and researchers in the context of a community of inquiring into mathematics teaching practices. In H. Sakonidis & D. Desli (Eds.), *Proceedings of the 2nd GARME* (pp. 556–567). Athens: Typothito. [in Greek].
Sierpinska, A., & Lerman, S. (1996). Epistemologies of mathematics and mathematics education. In A. J. Bishop et al. (Eds.), *International handbook of mathematics education* (pp. 827–876). Dordrecht: Kluwer Academic.
Stathopoulou, C., & Kalabasis, F. (2006). Language and culture in mathematics education: Reflections on observing a Romany class in a Greek school. *Educational Studies in Mathematics, 64*(2), 231–238.
Steinbring, H. (2006). What makes a sign a mathematical sign?—An epistemological perspective on mathematical interaction. *Educational Studies in Mathematics, 61*(1/2), 133–162.
Tzekaki, M., Kaldrimidou, M., & Sakonidis, H. (2002). Reflections on teachers' practices in dealing with pupils' mathematical errors. In J. Novotna (Ed.), *Proceedings of the 2nd CERME* (Vol. 1, pp. 322–332). Prague: Charles University.
Valero, P. (2004). Socio-political perspectives on mathematics education. In P. Valero & R. Zevenbergen (Eds.), *Researching the socio-political dimensions of mathematics education: Issues of power in theory and methodology* (pp. 5–23). Dordrecht: Kluwer Academic.
Walkerdine, V. (1998). *Counting girls out*. London: Falmer Press.
Wenger, E. (1998). *Communities of practice: Learning, meaning, and identity*. Cambridge, MA: Cambridge University Press.
Yackel, E., & Cobb, P. (1996). Sociomathematical norms, argumentation, and autonomy in mathematics. *Journal for Research in Mathematics Education, 27*(4), 458–477.

Chapter 6
A Speech Act in Mathematics Education – The Social Turn

David Wagner

When Stephen Lerman (2000) identified the social turn in mathematics education research, he described the phenomenon but also shaped it. He characterized the social turn as "the emergence into the mathematics education research community of theories that see meaning, thinking, and reasoning as products of social activity" (p. 23). In the tradition of the social turn, this chapter considers Lerman's claim in its social context by drawing from another theory that also emerged to consider its social context, namely speech act theory (e.g., Searle 1979) in the field of linguistics.

After introducing some fundamentals of speech act theory, in this chapter I illustrate them with mathematical examples. Next, I identify what Lerman claimed to be doing in the book chapter in which he identified the social turn and compare it to the way other scholars positioned his claim. From this, I reflect on the politics of Lerman's speech act and how it represents a model for other researchers. Like others, his speech act rests on authority. I then return to the mathematical examples of speech acts to identify similar politics in mathematics classrooms. My reflection closes with an alternative reading of the politics of his and other speech acts.

Speech Act Theory

Early speech act theory appealed to everyday examples of utterances to distinguish between a speaker's intention and the effects of the speaker's utterance. For example, utterances that have the grammatical structure of a statement, sometimes called declaratives, generally describe something. However, declaratives can do more than describe the world as it is. They can change the world. To exemplify the

D. Wagner (✉)
University of New Brunswick, Fredericton, NB, Canada
e-mail: dwagner@unb.ca

P. Gates, R. Jorgensen (Zevenbergen) (eds.), *Shifts in the Field of Mathematics Education*, DOI 10.1007/978-981-287-179-4_6

difficulty of appealing to grammar alone to identify an utterance as a declarative, Austin (1975) pointed to bequests, bets, the performance of marriage ceremonies, and the naming of ships. When the authorized person says, "I name this ship the Queen Elizabeth" (p. 5), it is the utterance itself that christens (names) the ship. Similarly a designate of the state changes the status of people by pronouncing them married; it is the pronouncement itself that begins the formal union. Bets and bequests change the status of the speaker as well as the status of others. Austin describes such acts in this way: "The uttering of the words is, indeed, usually a, or even the, leading incident in the performance of the act, [. . .] the performance of which is also the object of the utterance" (p. 8, emphasis his). However, he also recognized that the speech act is only consummated when the context aligns with the utterance: "[I]t is always *necessary* that the circumstances in which the words are uttered should be in some way, or ways, *appropriate*" (p. 8, emphases his).

Speech act theory, with its careful attention to the effects of language on social situations, has been influential in more recent theorizations of language and social dynamics. For example, Halliday's (e.g., 1978) Systemic Functional Linguistics distinguishes between a speaker's intention (illocutionary force) and the effects of the speaker's utterance (perlocutionary force) by analysing the utterance itself (locutionary force). Halliday, a linguist, noted how language in mathematics is developed to increase "its range of social functions" (p. 195). Halliday's theory has been used widely in mathematics education research, starting with Pimm (1987). Positioning theory also built on the idea from speech act theory that it is important to consider social situations. Positioning theory, as developed by social psychologists Harré and van Langenhove (1999), went beyond conventional psychology and attended to people's word choice in social situations and how that word choice initiates and sustains the roles of people in their relationships. In mathematics education research it is common to describe the 'positioning' within interactions, but such interpretation is often done without referencing positioning theory, as I have noted elsewhere (Wagner and Herbel-Eisenmann 2009).

Speech Acts in Mathematics

Special instances of speech acts occur in mathematics and mathematics classroom discourses. Like the non-mathematical examples listed above, mathematical actions involve naming and reformulating relationships. The nature of naming is especially relevant to Lerman's speech act.

When the owner of a steamship line christened a ship the Queen Elizabeth, people started using the new name to represent this particular ship. A student or mathematician may name (or christen) a quantity in the same way. When I work on a problem involving handshakes, for example, I might say, "Let x represent the number of people in the room." Before I make my statement, x does not represent the number of people in the (hypothetical or real) room. My statement turns x into an index of that quantity. Similarly, to solve the equation $e^{2x} - 10 = 3e^x$, I would find it useful to say (to myself or to others) "Let $y = e^x$" because this speech act turns a fairly messy equation into a simple quadratic equation, $y^2 - 10 = 3y$. In geometry, similar speech acts are useful. Given a diagram, it is useful to label key points with letters. When I put labels on the vertices of a triangle, for example, I christen these points as A, B, and C, and thus facilitate the communication that makes certain work possible. This kind of indexing, a special form of naming, is a powerful tool in mathematics for facilitating the development of ideas.

Netz (1998) analysed the role of diagrams in early Greek geometry and noted that "[p]oints are assigned letters – they are baptised" (p. 34) (baptism is associated with christening/naming in some cultures). He noted the significance of the order in which one represents ideas. Sometimes the mathematical object is determined first by text and then with a diagram, followed by the naming of certain points. Other points initially enter into play from the diagram. Either way, this suggests that the diagram itself, like words, is an utterance or text that represents some sort of mathematical idea that exists first in one's imagination. Radford (2002) theorized this mathematical necessity, in which people use spoken or written symbols to refer to objects that have no concrete existence. Pointing things out by naming them is a speech act because the act draws attention to something in particular and gives that thing a signifier to facilitate communication about it. Radford used the linguistics term 'deixis' to refer to such pointing (the root of 'deixis' is the Latin word for finger) and coined the term 'objectifying deixis' to describe the "process that makes apparent something new" (p. 18). Just as a ship exists as a concrete thing before its owner christens it, the mathematical examples of indexing I gave above involve naming ideas and objects that were already manifest in the world. By contrast, with objectifying deixis, a mathematical object is christened into being because it had no concrete form before it was named.

Kalthoff and Roehl (2011) pointed to a related phenomenon in mathematics education that might be seen as an opposite to Radford's objectifying deixis. Mathematics teachers (and, I would add, applied mathematicians) impose mathematical objects that exist in their minds onto everyday objects that are not as perfect or pure as the ideal object. For example, if we use a chocolate bar to talk about triangular prisms, we must ignore the aspects of that chocolate bar that make it less than a perfect prism and we must ignore the distractions of colour, smell, and taste. Kalthoff and Roehl described the performative nature of speech that turns the less-than-perfect object into an image of a perfect prism. For example, I could hold up a chocolate bar and say, "This is a triangular prism." It is not a perfect prism, but my speech act allows the people I am talking with to use the physical thing as an index for a perfect prism. This is a form of abstraction achieved by describing an object as though it is a perfect mathematical object (instead of a perfect chocolate bar, for example). Like indexing and objectifying deixis, these acts of abstraction are powerful because the abstraction makes possible certain kinds of calculations that would be otherwise cumbersome.

As noted by Halliday (1978) and others, nominalization, which is the process of making nouns or names for all sorts of things, is an important feature of mathematical language practices: "[L]ocutions with nominals in them have a greater semantic and syntactic potential for different emphases and different information structures" (p. 202). The speech acts I described above are all examples of nominalization.

In addition to acts of naming/nominalization, there are mathematical moves that are parallel to the declaration of marriage by a justice of the peace. At my wedding, the minister cleverly referenced my mathematics background by expounding on the nature of unions. He noted that he had often heard people say that $1 + 1 = 1$ in marriage because the two become one, but he argued that in marriage $1 + 1 = 3$; my partner and I would each keep and develop our own identity and also develop a new identity together. No one was fooled by his math. We all knew that two independent bodies do not magically become one body or three bodies, but we recognized the wisdom (or power) in thinking about things that are separate as though they are together. Mathematical operations are similar. For example, $2 + 3 = 5$. A set of two objects and a set of three objects may be seen as a set of five objects. This move to connect two groups of objects can facilitate certain actions that are difficult if we continue to separate the two objects from the three objects. It is the addition statement that changes the way we see the various objects with which we are working. We are not fooled into thinking that something magical happened with the objects. The only thing that changes is the way we think of the objects. However, we recognize the wisdom (or power) in such shifts of attention.

The above examples are relatively local mathematical acts performed by utterances but there are larger scale speech acts too. For example, there have been arguments about terminology for the roots of polynomial equations that do not appear on a number line. Gauss (1863, p. 177) complained about Descartes dismissively calling such numbers imaginary:

> That the subject [of imaginary magnitudes] has been treated from such an erroneous point of view and enveloped with such mysterious obscurity is due largely to the inadequate terminology used. If instead of calling +1, -1, $\sqrt{-1}$ the positive, negative and imaginary (sometimes even impossible) unities, they had been called, say, the direct, indirect and lateral unities, this obscurity would have been avoided. (quoted in Dantzig 1930/2005, p. 243)

Perhaps Descartes did not intend to christen $\sqrt{-1}$ as imaginary, but his term stuck and continues to be in force.

A Speech Act in Mathematics Education

Given the above description of speech acts and examples from mathematics discourses, I return to my claim that Stephen Lerman's identification of the social turn in mathematics education was a speech act. The proof that it was a speech act is that the declaration had an impact on the community in which it was spoken. With his declaration, Lerman described a phenomenon but also shaped the phenomenon by naming it. Perhaps his speech act was a case of objectifying deixis, using

Radford's (2002) term—a case in which Lerman made something that was vague and not concrete in itself into something that can be discussed and worked with. There was (and always is) movement and development within the field of mathematics education, but Lerman brought to our attention this particular development and gave us a way of talking about it.

What Did Lerman Claim About the Social Turn?

Before looking at the way others have written about the turn that Lerman pointed out, I will consider how he positioned his own speech act. His identification of the social turn was no doubt partially an intuitive recognition of a trend in theoretical frameworks, but it also built on careful analysis of more than a decade of "papers from *Proceedings of the International Group for the Psychology of Mathematics Education* (PME), from *Educational Studies in Mathematics* (ESM) and from the *Journal for Research in Mathematics Education* (JRME)" (Lerman et al. 2002), which Lerman reported on in various contexts with his collaborators. In their reporting they noted that an "orientation towards social theories of one kind or another is increasing" (p. 37).

The first publication in which Lerman referred to "the social turn" was a chapter in the book edited by Jo Boaler (2000), *Multiple Perspectives on Mathematics Teaching and Learning*. The title of Lerman's (2000) chapter is "The social turn in mathematics education research." This is how he introduced the social turn:

> I have called these developments the social turn in mathematics education research. This is not to imply that other theories, mathematical, Piagetian, radical constructivist, or philosophical have ignored social factors [. . .]. The social turn is intended to signal something different; namely, the emergence into the mathematics education research community of theories that see meaning, thinking, and reasoning as products of social activity. (p. 23)

As with most declaratives, when Lerman identified the social turn he described what was happening in mathematics education at the time. However, Lerman seemed to be aware of the significance of his act when he used the past tense to recognize his agency in identifying the trend. He chose to write, "I have called these developments the social turn" (p. 23), but he could have obscured his agency in various ways. For example, he could have used a passive voice to say, "There has been a social turn" or he could have foregrounded the agency of the community of researchers by saying, "mathematics education researchers have made a social turn." Lerman also showed awareness of his act of agency by writing about intention – "the social turn is intended to signal something different" – though he could have taken even greater ownership by saying, "I intend to signal something different."

Taken as a whole, Lerman's chapter demonstrates self-awareness of his positioning. He positioned himself not only as a reporter on trends but as an advocate for the shift he was reporting on. In his conclusion he focused on the difference between socio-cultural theory and psychological theory in the context of the unit of analysis in research: “But the object of study itself needs to take account of all the dimensions of human life, not a fragment such as cognition, or emotion” (p. 37). By using the imperative ‘needs to’, he christened the social turn into reality.

What Have Others Said About the Social Turn?

According to positioning theory, it is necessary to consider the reciprocal nature of positioning (van Langenhove and Harré 1999), which refers to the way an utterance is positioned by the speaker and by others in the context. While it is important to see how Lerman positioned his identification of the social turn, it is equally significant how others have positioned his speech act. The first to position his identification of the social turn was Boaler (2000, p. 6). Hers was a unique positioning because of her role as editor of the book in which Lerman declared the turn; it is like she was standing behind or beside Lerman when he made his utterance. She wrote:

> In an interesting analytical move he raises the importance of sociological theories to account for [power relations that are differently distributed across learners] — thus acknowledging the individual differences within social accounts of learning and employing sociological analyses to account for broader patterns of difference across individuals.

Boaler attributed agency to Lerman’s act. She did not say that he identified or recognized the social turn. Instead, she wrote that he raised the importance of sociological theories.

How did Lerman’s speech act raise the importance of sociological theories? How did his act make things possible that were not as possible before? These questions are central to my calling Lerman’s utterance a speech act. To answer this, I look to instances in which others have referred to his identification of the social turn, although this approach merely scratches the surface of what Lerman did. While I am sure that his speech act influenced the research behind publications that do not explicitly cite his claim, I limit myself in this chapter to explicit references. I can identify three things that his move has supported for others: attribution shields, further steps in the same direction, and critique.

First, attribution shields are prevalent in research reporting; most citations are examples of attribution shields. Linguistics uses the term ‘hedges’ to describe how writers and speakers make language intentionally fuzzy. An attribution shield is a particular kind of hedge, identified by Prince et al. (1982), in which one avoids providing rationale by attributing an idea to someone else who is deemed an authority. This kind of attribution is useful in research reporting because it allows us to focus on new ideas by avoiding discussion of ideas that have already been established.

For example, Gutstein (2003) noted that a reason for social justice issues not being associated with mathematics education was that “researchers have historically focused more on cognition than on sociocultural contexts (although this is changing)" (p. 41). He attributed this claim to Lerman. Gutstein was suggesting that Lerman's research is taken-as-shared and thus not up for argument. This move allowed Gutstein to focus on the new research he was contributing. In this way, Lerman's research, like most good research, provided a basis for others to move the field forward by building on his work. There are numerous examples of scholars who used Lerman's identification of the social turn in this way.

Second, I have identified instances of others taking what I call "further steps" to characterize the field of mathematics education and articulate imperatives for it. For example, Gutierrez (2010) and Valero (2004) have argued separately for a "socio-political turn" in mathematics education in much the same way as Lerman argued for the social turn; they all identified a turn and also advocated for it. Valero (2004) argued that Lerman's sense of the word 'social' was more encompassing than a straightforward reference to social theories. She highlighted his suggestion "that some researchers in mathematics education started focusing on the fact that there seemed to be a systematic exclusion of some students from the possibility of engaging in the learning of mathematics" (p. 12). She thus argued that this political concern was indicative of a "political turn" and to exemplify this she cited work as far back as fifteen years earlier, when Mellin-Olsen (1987) wrote *The Politics of Mathematics Education.*

Similarly, Gutierrez (2010) recognized that "Lerman's meaning of the term 'social' went beyond the layman's [sic] definition of involving social beings and interactions and included the consequences for addressing hegemony in society" (p. 4). However, Gutierrez positioned her claim differently from Lerman and Valero, who implied that the field was taking a turn. By contrast, Gutierrez (2010) suggested a split in the field: "[W]hile many mathematics educators are comfortable with including social and cultural aspects in their work, most are not so willing to acknowledge that teaching and learning mathematics are not politically neutral activities" (p. 4). Nevertheless, Lerman's speech act was a model for other researchers who sought to characterize and advocate for changes in the field.

Third, there are critiques of Lerman's identification of the social turn. These critiques have not said that Lerman's claim was unfounded, but rather that the social turn is problematic. Pais and Valero (2012) opened the social and political turns to question by referring to them as the "so-called" social and political turns, and then claimed that the identification of the social and political turns in mathematics education had missed the mark for reflexivity, which would require questioning the object of study. Stinson and Bullock (2012) took up this criticism and explained it with the metaphor of zooming out; the research focus remains the same but the social and political turns help us see more of the surroundings. Even with these turns, the focus remained on the "agenda that primarily explores questions of how to improve mathematics teaching and learning" (p. 45). They advocated for critique of this object of inquiry. Lerman's speech act made possible such critique. Unless researchers identify the movement (turn) in the field of mathematics education, it is difficult to ask what this movement does and does not do.

The Politics of Lerman's Claim

Just as Lerman argued for the necessity of interrogating mathematics teaching and learning in its sociocultural context, it is appropriate to zoom out and look at Lerman's claim in the sociocultural context of mathematics education. And, as suggested by the scholars who have upped Lerman's claim to include political aspects, it is appropriate to raise political questions—in this case to raise these questions about his claim in its context. What were the power relations at work in Lerman's act of naming the shift in attention to sociocultural realities? In this section I identify some uncomfortable aspects of Lerman's speech act. Although these may feel out of place in a book celebrating his work, I argue that any powerful act is inherently complex because of its social dynamics. The complexity of the political milieu is a testament to the power and necessity of Lerman's work.

For Lerman's speech act to have weight it is necessary that his authority be recognized within the mathematics education community. To illustrate the necessity of this authority, I return to the example of the naming of the ship that was eventually called the Queen Elizabeth. Any pet names, joking names, and functional/descriptive names that would have been used to refer to the ship during production would have had no staying power once the ship had been christened. Also, if someone had snuck into the shipyard at night before the christening and said, "I christen this ship the Queen of the Sea," the name would have had no power because the person would not have been acting in authority. For the christening to be accepted at large, it had to be enacted by someone with the authority to name the ship once and for all. Furthermore, the person in authority had to speak within a context that engaged his or her authority. For example, if the same person who officially christened the ship had snuck into the shipyard at night to christen the ship, that secret christening would not have had power. Thus two criteria are necessary for a speech act to succeed: the speaker needs to have authority and the context must invite an authoritative speech act.

These two criteria were in force for Lerman's speech act: the right context and the right person. In his public declaration of the social turn, Lerman (2000) used the past tense to say that he had "called these developments the social turn" (p. 23). This suggests that he had identified the turn before. However, when he did this on his own or among colleagues, it did not have the same force as when it appeared in a prominent mathematics education publication, which was the right context for a global speech act. Significant speech acts in the academy usually happen in publications. When these acts occur in different ways, it can be difficult for researchers to cite the origins of an idea. For example, I understand that although Ubiritan D'Ambrosio was talking about ethnomathematics at conferences in the early 1980s, he did not use the word in an English publication (the journal *for the learning of mathematics*) until 1985. Some scholars point to the 1985 publication as the moment in which D'Ambrosio coined the term, but others are aware that it was relatively well-known before then.

Lerman's authority was also necessary for the success of his speech act. He based the claim on a careful analysis of mathematics education literature. This was substantiated by his authoritative position in the community. Lerman had been a member of the International Committee and president of the International Group for the Psychology of Mathematics Education a few years before his speech act was published. He was also influential in the newly-formed Mathematics Education and Society group, which had its first conference in 1998. I can only speculate about the effect of the declaration of the social turn if a relatively unknown scholar were to have identified it. Would other scholars have paid attention in the same way? My graduate and undergraduate students often invent terminology to describe a phenomenon they notice in the literature but this does not have the same power as someone with Lerman's stature in the community identifying a movement within the field.

The recognition of Lerman's authority in our field raises the question of how he developed this authority. I suggest that his authority came from two forms of activity in the field: service and scholarship. While his leadership in key organizations exemplifies his service, his numerous acts of informal service cannot be ignored. Indeed, the small acts likely underpinned the trust others placed in him in his leadership roles. I did not know him at the time of his 2000 speech act, but since then I have observed him paying close attention to the work of both novice and experienced scholars and raising questions that support further development.

His scholarly activity complemented his service. What fascinates me is that Lerman's authority to make a powerful speech act rested to some extent on the fact that he had made other powerful statements before. His numerous publications bore many examples of speech acts before 2000. In those publications he identified phenomena and provided descriptions that indexed ideas, and thus moved the field forward. A person develops the authority to move others with words by saying powerful things that become increasingly recognized in that person's community.

Though we hold our academic colleagues in esteem for their activity in the field, it is worth raising questions about this activity. As I noted above, authority is central to speech acts like Lerman's. In Judeo-Christian and other cultures, giving names is associated with power. In a Jewish creation story, Adam (the first human) is authorized by God to name the animals, and, in the same breath, to rule over them (exercise dominion over them). Leguin (1988), indexes this story in a modern fiction piece that turns this dominion on its head. The character she called Man un-named the animals, which indicated a release of power. This release prompted Man to be more attentive to experience:

> I could not chatter away as I used to do, taking it all for granted. My words now must be as slow, as new, as single, as tentative as the steps I took going down the path away from the house, between the dark-branched, tall dancers motionless against the winter shining. (p. 196)

Thus, when we as scholars draw attention to something, we are also drawing attention away from other things. Drawing on Leguin's insight, I would say that when I draw people's attention to something I distract them from noticing or attending to the experience themselves. This is like the abstraction I described above with the example of the perfect prism abstracted from a chocolate bar. Hermeneutical phenomenology is a research method that attempts to undo the directed attention that comes with language and names given by others. There are a few examples of such research in mathematics education, but I claim that the principle is warranted for any methodology. We need to ask what we are ignoring through our choices of objects of study. The critiques raised by Pais and Valero (2012) and by Stinson and Bullock (2012), noted above, are examples of scholars doing this for the field, but we can all do this as individual researchers too.

Colonized settings help make clear the power and dominion related to names. For example, the river that flows near my home is called by most people and labelled on maps as the Saint John River. The river's original name among the Indigenous people was the Wolastoq but English settlers renamed the river after a Christian saint. The fact that the English name has eclipsed the Maliseet name is an indication of historical and current power relations. It is likely that various settlers referred to this big river in various ways until someone seen to be an authority (probably Samuel de Champlain) declared it to be the Saint John River.

Similarly, when the mathematics education community heeds an important characterization of the field articulated by a respected person like Stephen Lerman, the community may ignore other characterizations of the field. Furthermore, I claim that European and North American scholars, particularly male and English-speaking scholars, have had the advantage of being the first ones active in the field because of power relations that permeate historical and current geopolitics. I am saying this not to discount the good work by Lerman and others but rather to argue that we need to be attentive to voices from the margins when they speak and write about our field.

As I noted in the mathematical examples above, there are local and global speech acts. Lerman wrote that he had identified the social turn and called it that some time before he introduced it to the global mathematics community. There is value in scholars characterizing the field for themselves, but for the development of the field the political questions become prominent.

The Politics of Mathematical Speech Acts

Because I used mathematical examples to illustrate speech act theory, and because our scholarly community is characterized by our interest in the teaching and learning of mathematics, I will return to the examples of speech acts in mathematics. I noted above how speech acts can be useful in developing mathematical ideas. Now, having considered the power relations at work with speech acts in scholarship, I note that similar dynamics are at work in the development of mathematics both for learners and for mathematicians. I will focus on learners. The potential power of mathematical speech acts is mediated by the power relations in their contexts.

For example, if a group of students is working on a mathematical problem, certain people in that group are better positioned than others to perform objectifying deictics, to index values or points, and to abstract ideas. When they label points on a diagram, their labels stick better than when others try to label the same points. When they suggest the introduction of a variable, it carries more force than when others do so. When they see a particular object as an example of a perfect mathematical idea, it is accepted more than if others did so. These students may acquire this power through a history of initiating similarly powerful mathematical ideas. But other factors are at play too—including gender, race, and other identity-related dynamics. Similar dynamics are at work in whole-class discussion, in which the teacher alone is too often positioned as the only one capable of powerful mathematical speech acts.

One reason that I value the teaching of mathematics is that it can equip people to use mathematics in society. In this way people have access to a discourse in which sound reasoning accompanied by clear explanation can trump status hierarchies. Thus mathematics has the potential to help people challenge inequities in society. However, if access to mathematics in schools is compromised by status hierarchies in classrooms, some of the children who might use mathematics well in society could be discouraged from seeing themselves as capable of using mathematics. Indeed, as I noted above, Valero (2004) made the case that the problem of access to powerful mathematics was explicitly cited by Lerman in his justification for attention to the social in mathematics education research.

Power and Intimacy

Though naming is associated with power, it is more than power; it also represents intimacy and knowledge. Thus I will close with an alternative reading of the politics of Lerman's speech act. Names associate with stories and experience. Van Manen et al. (2007) described the significance of naming in the context of personal names: "The stories of who named us and why that particular name was chosen are a link to our origin and take on significant meaning for us" (p. 85). They illustrated both the intimacy associated with the knowledge and use of a correct name and the profanity of using a name carelessly or incorrectly. The prohibition against the use of God's name in vain, as referenced in the Ten Commandments, is an extreme case of this abuse in the Judeo-Christian tradition.

Authority can rest on intimacy as much as it does on power. I might even refer to intimacy as a form of well-earned relational power. To illustrate, I may be receptive to critique or guidance from anyone, but I am most receptive to critique and guidance from someone who knows me well and who is committed to a relationship with me. In short, their intimacy with me allows them to move me to action or change.

I close this chapter by foregrounding this aspect of Stephen Lerman. He has intimate knowledge of the field of mathematics education and he has shown his commitment to the field and its people. A duty of such intimacy is to critique the field, just as a good friend might offer a difficult but well-meaning word of advice or guidance. I have noted that Lerman has had some advantages over non-native English speakers, for example, but he has also worked very hard and conscientiously. Most importantly, he has used the positions that he has found himself in and that he has made for himself to amplify the voices of people at the margins both in mathematics classrooms and in the mathematics research community. For this, he has garnered my deep respect.

References

Austin, J. (1975). *How to do things with words*. Cambridge, MA: Harvard University Press.

Boaler, J. (2000). Introduction: Intricacies of knowledge, practice, and theory. In J. Boaler (Ed.), *Multiple perspectives on mathematics teaching and learning* (pp. 1–18). Westport: Ablex.

D'Ambrosio, U. (1985). Ethnomathematics and its place in the history and pedagogy of mathematics. *For the Learning of Mathematics, 5*(1), 44–48.

Dantzig, T. (1930/2005). *Number: The language of science*. New York: Plume.

Gauss, C. (1863). *Theoria residiorum biquadraticorum Commentatio secunda*; Werke, Bd. 2. Goettingen.

Gutierrez, R. (2010). Editorial: The sociopolitical turn in mathematics education. *Journal for Research in Mathematics Education, 41*, 1–32.

Gutstein, E. (2003). Teaching and learning mathematics for social justice in an urban, Latino school. *Journal for Research in Mathematics Education, 34*(1), 37–73.

Halliday, M. (1978). *Language as social semiotic: The social interpretation of language and meaning*. Baltimore: University Park Press.

Harré, R., & van Langenhove, L. (1999). *Positioning theory: Moral contexts of intentional action*. Oxford: Blackwell.

Kalthoff, H., & Roehl, T. (2011). Interobjectivity and interactivity: Material objects and discourse in class. *Human Studies, 34*, 451–469.

Leguin, U. (1988). *Buffalo gals and other animal presences*. New York: Penguin.

Lerman, S. (2000). The social turn in mathematics education research. In J. Boaler (Ed.), *Multiple perspectives on mathematics teaching and learning* (pp. 19–44). Westport: Ablex.

Lerman, S., Xu, G., & Tsatsaroni, A. (2002). Developing theories of mathematics education research: The ESM story. *Educational Studies in Mathematics, 51*, 23–40.

Mellin-Olsen, S. (1987). *The politics of mathematics education*. Dordrecht: Kluwer.

Netz, R. (1998). Greek mathematical diagrams: Their use and their meaning. *For the Learning of Mathematics, 18*(3), 33–39.

Pais, A., & Valero, P. (2012). Researching research: Mathematics education in the political. *Educational Studies in Mathematics, 80*, 9–24.

Pimm, D. (1987). *Speaking mathematically*. London/New York: Routledge and Kegan Paul.

Prince, E., Frader, J., & Bosk, C. (1982). On hedging in physician-physician discourse. In R. J. Di Pietro (Ed.), *Linguistics and the professions. Proceedings of the second annual Delaware symposium on language studies* (pp. 83–97). Norwood: Ablex.

Radford, L. (2002). The seen, the spoken and the written: A semiotic approach to the problem of objectification of mathematical knowledge. *For the Learning of Mathematics, 22*(2), 14–23.

Searle, J. (1979). *Expression and meaning: Studies in the theory of speech acts*. New York: Cambridge University Press.

Stinson, D., & Bullock, E. (2012). Critical postmodern theory in mathematics education research: A praxis of uncertainty. *Educational Studies in Mathematics, 80*, 41–55.

Valero, P. (2004). Socio-political perspectives on mathematics education. In P. Valero & R. Zevenbergen (Eds.), *Researching the socio-political dimensions of mathematics education: Issues of power in theory and methodology* (pp. 5–24). Dordrecht: Kluwer.

van Langenhove, L., & Harré, R. (1999). Introducing positioning theory. In R. Harré & L. van Lagenhove (Eds.), *Positioning theory: Moral contexts of intentional action* (pp. 14–31). Oxford: Blackwell.

Van Manen, M., McClelland, J., & Pilhal, J. (2007). Naming student experiences and experiencing student naming. In D. Thiessen & A. Cook-Sather (Eds.), *International handbook of student experience in elementary and secondary school* (pp. 85–98). New York: Springer.

Wagner, D., & Herbel-Eisenmann, B. (2009). Re-mythologizing mathematics through attention to classroom positioning. *Educational Studies in Mathematics, 72*(1), 1–15.

Part III
Steve Photo Selection

Chapter 7
Steve Through the Years

Peter Gates and Robyn Jorgensen (Zevenbergen)

Fig. 7.1 Steve thinking about his career age 4

P. Gates (✉)
School of Education, University of Nottingham, Wollaton Road, Jubilee Campus, NG8 1BB
Nottingham, UK
e-mail: peter.gates@nottingham.ac.uk

R. Jorgensen (Zevenbergen)
University of Canberra, Canberra, Australia
e-mail: Robyn.Jorgensen@canberra.edu.au

P. Gates, R. Jorgensen (Zevenbergen) (eds.), *Shifts in the Field of Mathematics Education*, DOI 10.1007/978-981-287-179-4_7

Fig. 7.2 Steve practising public speaking to brother Tony

Fig. 7.3 Steve with his mother, father and brothers Michael and Tony

Fig. 7.4 Steve in the early 1970s

Fig. 7.5 Steve and his daughters in Paris

Fig. 7.6 Steve and family in Spain for chanucah

Fig. 7.7 Steve and Beryl in the early 1980s

Fig. 7.8 Steve Lerman, Paul Ernest, Charalampos Sakonidis PME Hungary 1988

Fig. 7.9 Steve and daughters in Hungary

Fig. 7.10 Steve and daughters in Vienna on route to Hungary

Fig. 7.11 PME Mexico 1990

Fig. 7.12 Steve at his PhD ceremony Albert Hall

Fig. 7.13 Steve Lerman, Charalampos Sakonidis and Marianna Tzekaki in Cape Town 1998

Fig. 7.14 At a Wittswatersrand graduation

Fig. 7.15 PME in Hawaii July 2003

Fig. 7.16 Steve with daughters Rebecca and Abigail

Fig. 7.17 Steve presenting at PME30 Prague 2006

Fig. 7.18 Steve and Beryl

Fig. 7.19 Steve in 2014 in France

Fig. 7.20 Steve in the Wittswatersrand 2014

Fig. 7.21 Steve in Australia 2014

Fig. 7.22 Steve with Beryl in Australia 2014

Fig. 7.23 Steve Lerman with Peter Gates at PME 38 Vancouver July 2014

Fig. 7.24 Steve in Grahamstown writing workshop July 2014

Part IV
Steve and International Cooperation

Chapter 8
International Research Collaborations: An Australian Perspective

Robyn Jorgensen (Zevenbergen)

In the Australian research context, collaboration with international researchers is highly valued. In part, this is due to the geographical isolation of the nation and the desire to ensure that the research that emanates from the country is at world-class level. Working with international researchers who are well respected within their communities is one of the pillars upon which many Australian universities build their strategic plans. This chapter explores the international collaboration between Steve and myself.

The story that I present in this chapter is not uniquely an Australian collaboration. Steve has worked with many people across many nations. There are many examples of his collaboration on international projects that are included in this collection of work. These include but not restricted to his work in South Africa on teacher education with Jill Adler, his work with Joao Filipe Matos and a Portuguese team, exploring the newspaper selling practices in the eastern cape of Africa, building theoretical works with Anna Tsatsaroni or working with David Clarke and a multitude of international researchers on the Lesson Study project. Anyone who has worked with Steve over a long project will attest, it is both a pleasure and professionally enriching experience. While this chapter explores the more formal research context of the competitive research grant system that operates in Australia, I also acknowledge the international collaborations that Steve and I have undertaken with Jo Boaler in the US. We had the pleasure and privilege of working with Jo on her well known work in Railside. In this work we drew on our three different theoretical models to frame the practices that were observed in the Railside context. It is certainly a rare academic who has worked with so many people in so many contexts across so many projects and all those who have worked with him still hold him in great esteem. This is testament to a very unique academic. It is with a sense

R. Jorgensen (Zevenbergen) (✉)
University of Canberra, Canberra, Australia
e-mail: Robyn.Jorgensen@canberra.edu.au

P. Gates, R. Jorgensen (Zevenbergen) (eds.), *Shifts in the Field of Mathematics Education*, DOI 10.1007/978-981-287-179-4_8

of humility that I write this chapter and pay tribute to the amazing work Steve contributed to the various projects and to the pleasure of working with him.

Since first meeting Steve in 1992, we have worked collaboratively on a number of projects. Of these, there are two significant projects, both funded by the Australian Research Council in which Steve was a Partner Investigator (PI). To be a recipient of Australian Research Council grants in Australia is the peak of a researcher's career. These are the "gold-star" standard of research within Australia. This chapter outlines the integral and important role Steve had in shaping the grant applications and research outcomes. Steve's influence in these projects is very recognisable and reflected in the orientation to research and theory building.

I first met Steve when I was undertaking doctoral studies at Deakin University. I was trying to understand how the practices used by teachers of mathematics were implicated in the construction of social differences – a focus that has remained in my work since the 1990s. Steve had come to Deakin to participate in a conference on constructivism as the keynote speaker. It was about this time that he edited a special issue of *Educational Studies in Mathematics* on the theme of constructivism. The Deakin conference was the launching pad for what has become a 20-year-plus collaboration. Over the years, we would meet at conferences and talk through ideas, usually over a wine or two and a meal – usually fish or vegetarian. We also co-ordinated visits to Australia where Steve worked at Deakin (Geelong); Griffith (Gold Coast and Brisbane) and Charles Sturt (Albury) Universities and is currently an adjunct professor in the Griffith Institute for Educational Research. Our friendship and collegiality grew as did my respect for his work and his capacity to continually draw on new ideas to frame his thinking.

Steve's perspectives have played an important role in the formation and on-going progress in two large research projects that have been conducted in Australia with implications internationally. Negotiating our particular interests into the formation of innovative projects has been productive. The first of these projects was based on the notion that computers and other digital devices had become the focus of reforms in education but mathematics education was also caught in somewhat of a paradox. The potentiality of these new tools to enhance learning was undeniable yet the field was still debating the merits of the (relatively speaking) primitive device known as the calculator. It was also known that access to digital tools (mainly computers) was also split along particular social lines. At that time, the "digital divide" was common in the educational debates around their use (and access) in classrooms. The second of the projects was one in which we used a design research approach to work with teachers and communities to enhance the mathematics learning for Indigenous students. Each of these will be discussed in subsequent sections of this chapter.

The Australian Research Council: Background

To understand the importance of Steve's work on these collaborations, some background into the research granting schemes will provide a context for this chapter. The Australian Research Council (ARC) is the peak funding body for research in Australia. The grants awarded by the ARC hold considerable status for recipients and their employing Universities. There are two main schemes – *Discovery Projects* and *Linkage Projects*. The Discovery Scheme is aimed at blue sky research while the Linkage Projects are strongly grounded with Industry partnerships and aimed at solving practical problems. Both projects can be undertaken by individuals or teams of researchers: Linkage Projects must also be undertaken in collaboration with Industry Partners.

The Discovery Projects are underpinned by a number of key objectives that have remained relatively consistent, and are unsurprising for projects of this calibre. These objectives, as listed on the ARC website are:

- support excellent basic and applied research by individuals and teams;
- enhance the scale and focus of research in the National Research Priorities;
- expand Australia's knowledge base and research capability;
- encourage research and research training in high-quality research environments;
- enhance international collaboration in research;
- foster the international competitiveness of Australian research. (Australian Research Council 2012)

In contrast, the Linkage Scheme is more grounded and requires partnerships between researchers and industry. These partnerships are not only about the research to be undertaken but are strongly shaped by the commitment of the industry to the research. The ARC requires the industry partner to make a substantial contribution to the research through financial and in-kind support. For example, the 2012 rules for 2013 funding required industry to "make a significant contribution in cash and/or in kind support that is equal to, or greater than, the ARC funding". While having changed over the years, this has always been a feature of Linkage grants. The objectives of the Linkage scheme are somewhat different from, though inclusive of quality research, as the Discovery scheme. These objectives are to:

- initiate and/or develop long-term strategic research alliances between higher education organisations and other organisations, including industry and end-users, in order to apply advanced knowledge to problems and/or to provide opportunities to obtain national economic, social or cultural benefits;
- enhance the scale and focus of research in National Research Priorities;
- foster opportunities for researchers to pursue internationally competitive research in collaboration with organisations outside the higher education sector, targeting those who have demonstrated a clear commitment to high-quality research; and

- produce a national pool of world-class researchers to meet the needs of the broader Australian innovation system. (Australian Research Council 2012)

The success rate for applicants is relatively low. Depending on the scheme and the year, it is mostly around 20 %. Historically, the Linkage scheme initially had a higher success rate due, in part, to initial reviewing by industry as to the merits and applicability of the research. There has been a change in the recent rounds of Linkages where there are now similar success rates as for Discoveries.

Criteria

Over the years, the criteria and their relative weighting have evolved though there are elements that have remained relatively consistent. To highlight the differences in the two schemes, it is useful to consider the selection criteria against which the schemes are assessed. Panels are established for various disciplines. Members of these panels assess the projects along with experts in the field nominated by the Panel members. Up to five people outside the ARC College of Experts may assess any one project.

Discovery Projects

Discovery projects are assessed according to the guidelines set out by the ARC. For the 2011 round of application, the ARC (Australian Research Council 2011a, pp. 6–7) listed the following selection criteria and relative weightings

(a) Investigator(s) (40 %)
- Research opportunity and performance evidence (ROPE); and
- Capacity to undertake the proposed research.

(b) Project Quality (40 %)
- Does the research address a significant problem?
- Is the conceptual/theoretical framework innovative and original?
- Will the aims, concepts, methods and results advance knowledge?
- Are the project design and methods appropriate?
- Will the proposed research provide economic, environmental and/or social benefit to Australia?
- Does the project address National Research Priorities?

(c) Research Environment (20 %)
- Is there an existing, or developing, supportive and high quality research environment for this Project?

- Are the necessary facilities to complete the project available?
- Are there adequate strategies to encourage dissemination, commercialisation, if appropriate; and promotion of research outcomes?

Linkage Projects

The Funding Rules for 2012 Linkage projects (Australian Research Council 2011b, pp. 9–10) listed the criteria as follows:

Investigator(s) (20 %)
- Research opportunity and performance evidence (ROPE); and
- Capacity to undertake and manage the proposed research.

Proposed Project (50 %)
- Significance and Innovation (25 %)
  - Does the research address an important problem?
  - How will the anticipated outcomes advance the knowledge base?
  - Are the Project aims and concepts novel and innovative?
  - Will new methods or technologies be developed?
  - Will the proposed research provide economic, environmental and/or social benefit to Australia?
  - Does the Project address National Research Priorities?
- Approach and Training (15 %)
  - Are the conceptual framework, design, methods and analyses adequately developed, well integrated and appropriate to the aims of the Project?
  - Where relevant, is the intellectual content and scale of the work proposed appropriate to a higher degree by research?
  - How appropriate is the proposed budget?
- Research Environment (10 %)
  - Is there an existing, or developing, supportive and high quality research environment for this Project?
  - Are the necessary facilities to complete the Project available?

Nature of the Alliance and Commitment from a Partner Organisation (30 %)
- Is there evidence that each of the Partner Organisation(s) is genuinely committed to, and prepared to collaborate in, the research Project?
- Will the proposed research encourage and develop strategic research alliances between the higher education organisation(s) and other organisation(s)?
- Value for money and budget justification for cash and in-kind contributions.

While the criteria are quite different depending on the scheme, two features remain consistent across both schemes and over time. The project itself has to be of significance, of national benefit and be innovative to be funded. The investigators involved in the project carry significant weighting. Indeed, it is the case, that it is almost impossible to obtain a grant unless the researcher or team of researchers has a significant track record in the area of investigation. Where teams of researchers are applying for a grant, then the track record extends to the research team having collaboratively established a track record.

Numeracy, Equity and ICT: A Study of Classroom Practice

In early 2000, I spent 2 months of my first-ever sabbatical working with Steve at London South Bank University. During this time we were able to talk through our ideas for establishing a much stronger working relationship. It was during this time that we developed the groundwork for a competitive research grant – an Australian Research Council Discovery grant (See Lerman and Zevenbergen 2006, 2007). Over the next year or so, Steve commuted to Australia and we developed the project. In the final throes of the project writing and just prior to submission of the proposal, we included Peter Renshaw who was a new arrival at the Gold Coast campus and was keen to be involved in projects. This application was successful and funded for the period of 2003–2005 for an amount of AU$140,000 over the period of the project. While this amount may seem small in the international arena for educational research, the Australian grant system, unlike other contexts, is not full-cost recovery. The Universities are expected to provide the salaries for Chief investigators and university infrastructure. The funds are used for salaries of support staff such as research assistants and administrative or specialist support; travel; and equipment to enable this research project to be completed. Universities are expected to support the day-to-day costs associated with the operation of a project – such as office, telephones, postage, and so forth, as well as any tenured staff who are working on the project. Where the project may extend these resources – such as through the mail-out of a large survey – then the ARC would support this type of work as it deviates from the normal operational support of a university. Higher degree students who work on these projects are not funded by the projects unless specific funding has been nominated and gained for the project.

The project had four main aims (see Zevenbergen and Lerman, 2005, 2006, 2007, and 2008). These are taken from the application:

First, it aims to document the ways in which ICTs are used within schools, community contexts and homes to engage students in the middle years of schooling in numeracy learning, particularly students from disadvantaged regions and communities. Focused case studies will be conducted in a range of strategically-selected sites that involve participants of diverse social/cultural and geographical backgrounds. Drawing on the outcomes of national and international research that have documented the ways in which ICTs can enhance numeracy learning, the case-study sites will be evaluated for the affordances and constraints embedded in their current practices. Particular attention will be given to documenting the equity of access to ICT-mediated learning within and across sites.

Second, it aims to identify the new forms of numeracy that arise from ICT-mediated activities. The notion that numeracy itself is an evolving set of historically-situated social practices rather than a defined set of mental activities, arises from recent sociocultural theory. This project provides an opportunity, therefore, to further articulate sociocultural theory in relation to the nexus between numeracy, ICTs and equity in these "New Times".

Third, it aims to develop a pedagogical framework, relevant to the middle years of schooling, for ICT-mediated numeracy. Using the theory and analytical methods of Bernstein in conjunction with the sociocultural theory of Vygotsky, we propose to identify the characteristics of teaching practice that can enhance the effective use of ICTs for numeracy learning.

Fourth, it seeks to develop, implement and evaluate equity guidelines for teaching innovations that will enhance numeracy learning. We will focus here on those most at risk – indigenous students, students in remote or geographically isolated regions and working-class students. Gender will be integrated into these categories since this compounds the differential access and success in numeracy and ICT. The equity guidelines will provide a framework for teachers to determine whether their practices are inclusive and empowering.

To achieve these aims, schools from urban/regional, rural and remote areas were included in the study. The schools included targeted schools that had low-SES students and/or Indigenous students. This spread of schools was to ensure we would be able to understand the affordances and constraints of these sites on the use and uptake of ICT-mediated learning in mathematics. The study employed an action research methodology where teachers would come to a central location. Here they would experience input from the research team but even more importantly from each other. It was anticipated that schools could share their good practice with other colleagues so that a learning community would be facilitated. The schools were targeted as those who were using ICTs in their classrooms so it was seen to be a productive process. Classroom observations and videos were taken of lessons and then analysed using a framework that was developed in Queensland. This was framework – Productive Pedagogies (Education Queensland 2000) – allowed a rich analysis to be undertaken of the practices being employed by the teachers.

Steve's use of Bernstein was influential in informing the application and its carriage. The use of Bernstein enabled a rich way in which to analyse the ways the instructional discourse used by teachers either informed (or not) the enabling pedagogies of computer-based activities. Since there was a strong possibility that teachers may need to use new forms of pedagogy around the use of ICTs in the classroom, using Bernstein enabled a way of studying and theorising teachers' work. From the application, Steve's framing for the project is evident:

> The theoretical basis of this research is embedded in Bernstein's pedagogic discourse (1990, 1996). Bernstein claims that through different methods of teaching, students have greater or lesser access to the information being conveyed by teachers, depending on the extent to which the pedagogy is visible or invisible. Traditional teaching of mathematics has the advantage of being visible for all students, that is, the 'rules' for knowing what one is supposed to do in the mathematics classroom are made explicit to students. For example, the locus of authority in relation to knowledge is with the teacher and the textbooks, and this is known by students and teachers. More progressive pedagogies do not make these rules so explicit (Zevenbergen and Lerman 2001). The locus of control ostensibly shifts from the teacher to a shared authority amongst all the classroom participants but as a consequence many students, largely those from disadvantaged homes, do not read the mathematics tasks in the way that is expected and thus are not able to demonstrate their knowledge (Cooper and Dunne 1999). A further element of Bernstein's work that comes into play here is the boundary between the everyday and school mathematics. Contextualising mathematics questions can be misleading for students from disadvantaged homes. Drawing on these key ideas, and other notions concerning the effects of different pedagogies, the project seeks to identify theoretically and empirically aspects of pedagogy that are effective in terms of socially-disadvantaged students being able to access knowledge, skills and applications of learning technology in school mathematics.

The use of Activity Theory was also a strong framing for the project. This framework became a very useful and important part of the project. As the project evolved, it became increasingly clear that despite the time and professional learning that was being invested in the project, there were only marginal gains in terms of the uptake in the use of ICTs to support mathematics learning. It would have been easy to frame this lack of uptake by the project team members as being a deficit in some way. But through the use of Activity Theory, it became possible to frame this outcome in a far more positive light. Steve's influence and strengths in the use of Activity Theory helped to move the project into a richer framing of the outcomes through the use of Activity Theory and in moving away from deficit framing of the outcomes.

As part of the application process, there is a requirement for the authors to justify the inclusion of team members. As a condition of the application at that time, Steve (as a non-Australian) could assume the role of a Partner Investigator (PI). As part of the track record assessment, not only were Steve's research outputs important, but also his contribution to the project. Steve's role was listed as:

> **Prof Stephen Lerman** (PI) Lerman is an internationally renowned mathematics educator and is strongly placed in the area of sociocultural theories of learning. He has a particular interest in the work of Vygotsky and Bernstein and is involved both in a nationally funded project drawing on these ideas, and an internationally funded project. The latter is focused on analysing how classrooms produce mathematical thinking in students and in its first stage it has developed research techniques for analysing classrooms. Mathematical thinking is seen to be socially organised: different social groups are positioned differently and thus are liable to produce different forms of mathematical thinking. In the second stage empirical work is to be carried out in classrooms. These insights will be informative for this present project. Both Lerman and Zevenbergen have a strong research interest in the sociocultural practices of mathematics education.

It was also important for the assessment of the projects to illustrate collaboration among the research partners. This was relatively easy as Steve and I had collaborated on a number of writing tasks so it was clear that there had been joint work (and outputs) established.

The research project was successful in terms of the considerable number of publications produced. As the project progressed, the roll-out of interactive whiteboards was becoming more obvious across the schools involved in the project. As such, we included this element in the project as it was possible that this may have been a significant innovation in the use of digital media in the classroom. Using the same tools as we had been using with the adoption of computer-mediated learning, we analysed the teachers' use of interactive whiteboards. The publication of this work in the *Mathematics Education Research Journal* has resulted in it being in the top 10 downloaded papers of that journal.

Complex Instruction

The second major grant was somewhat longer in its gestation. The second project was based in remote Australia and with very low outcomes for learners in terms of literacy and mathematics. A number of events lead to the involvement of Steve in this project (See Sullivan et al. 2013).

In 2004, Jo Boaler invited Steve and me to work with her on her research in US schools. Boaler's work has been well documented, particularly the work around Railside (Boaler 2008; Boaler and Staples 2008). In the year prior to the visit, we drew on three different perspectives to analyse a snippet from a Railside classroom – Steve's work with Bernstein, my work with Bourdieu, and Jo's work with Complex Instruction –fertile ground for unpacking the classroom episode in order to understand how interactions could be enriched by different theoretical approaches. These were presented at a symposium at the American Educational Research Association's conference. What was integral to this meeting was a deeper understanding of Complex Instruction – a tool that enabled a more coherent approach to curriculum and pedagogy. Jo was able to promote this approach to us and through our lively discussions; we collectively gained a much richer understanding of ways in which schools were able to reform their teaching of mathematics, particularly for those students most at risk of failing. In the case of Railside, the school had gone from one of the poorest performing schools in the state, to above state average in a relatively short period of time. More interesting was that the reform impetus came from the mathematics department – usually one of the last bastions of change in a school.

Working with Jo and seeing the power of the reform was enabling and productive for us. A year later, Peter Sullivan and I were approached by the Association of Independent Schools in Western Australia (AISWA) at the national Mathematics Education Research Group of Australasia Conference to work with them on a project aimed at enhancing the mathematics learning for Indigenous students. This is an organisation that supports schools operating outside the state system. The schools in the remote Kimberley region of Western Australia worked together and were concerned with the learning, or more aptly, the lack of learning, of mathematics by their students.

There were six remote community schools that participated in the study. These are part of a cluster of Aboriginal schools, operated by the local Indigenous communities with a board that represents the community. Each community is a separate entity. The six schools were spread over a distance of about 1,000 km, many of them on dirt roads that could not be accessed in the wet season due to flooding. The schools wanted some support in building a strong mathematics program. The work the year earlier with Steve and Jo around Complex Instruction became the basis for successful grant application. The collaboration with Steve and Jo added strength to the application in terms of track record in this area.

As the team was in partnership with AISWA, the application fell into the Linkage Grant scheme. As the project had practical links and outcomes, involved working in context with a partner organisation which sought to have some input, it neatly fitted the requirements of the Linkage scheme (See Jorgensen et al. 2011).

Enhancing Mathematics Learning for Indigenous Students

The Kimberley schools were and still are, a very unique group of schools. Each of the six communities had taken control of the school was effectively operating independent of the government school sector. While still needing to be compliant with government requirements, each school was able to employ its own staff. The principal worked with, and was accountable to, the community board. As such, the principal acted as a conduit between the Council and the wider educational community. Each of the six Councils approved their school's involvement in the project.

As part of the application process, a summary of the proposal was written for the ARC to promote successful projects. This summary, reproduced below, provides a synopsis of what we proposed to undertake.

> Equity outcomes for Indigenous education are decreasing. This project seeks to implement high quality, high demanding mathematics in remote Indigenous communities in the Kimberley. The project recognises that learning mathematic demands a cultural approach for students who culture is not that of the school mathematics. Using a design research approach, the project explores quality learning environments for students, teachers and Aboriginal Education Workers. The project aims to develop sustainable practices in hard-to-staff regions that support high quality mathematics learning. The project will provide guidelines for the development of rigorous and culturally-appropriate practices in mathematics with application all equity contexts.

The aims for this project were to:

- Enhance the mathematics learning of Indigenous students living in remote areas of Australia through the provision of rich mathematical experiences. The focus of the project is the middle years of schooling as this provide a containable project while identifying the range of learning for a significant number of students;

- Support the mathematical content and pedagogical knowledge of teachers working with these student through the provision of professional learning communities. These communities will be both face-to-face and digital;
- Develop on-line support that provides sustainable learning for teachers working win remote communities;
- Develop professional learning for community representatives who remain in the communities with the explicit recognitions that these people have knowledge which is integral to the learning and sustainability of the communities.

For this grant, Steve was listed in the following way as the guidelines for this application required us to write "a statement of the most significant contributions to this research field":

> PI Lerman has an extensive career in mathematics education. He has worked on a number of ERSC project in the UK and from 2003–2005 worked with Zevenbergen on an ARC Discovery Grant as a PI. Stephen Lerman was President of the International Group for the Psychology of Mathematics Education from 1995–1998. This is the leading international research group in the field and holds annual conferences in different parts of the world. He was Chair of the British Society for Research in Learning Mathematics from 1994–1996. He is Senior Editor of the International Journal of Science and Mathematics Education, and is on the editorial boards for the following journals: Educational Studies in Mathematics, Journal of Mathematics Teacher Education, Mathematics Teacher Education and Development; Themes in Education: For the Learning of Mathematics; and Pythagoras.

As the Linkage Scheme focuses heavily on partnerships, we had to identify ways in which the team could collectively enhance the project. Again Steve's work was integral to this. We argued that the pedagogic relay which was drawn from Steve's early work in the late 1990s was important to theorising and understanding how the cultural aspects of school and that of the students may be implicated in the learning of mathematics for Indigenous students.

> CI Zevenbergen and PI Lerman have also had extensive experiences in process for teaching mathematics to students from disadvantaged backgrounds. Lerman (1988) attended to SES-related difference between classroom expectations and students' aspirations; and Zevenbergen and Lerman (2001) argued that decoding contextualised problems corresponds closely with students' SES backgrounds. Lerman also work with Bernsteinian theories (1996) to develop a sense of how the pedagogic relay is implicated in the access to mathematical learning. This theory has been developed in two earlier Discovery Grants in which the research team has been involved (DP044999) with Sullivan and Zevenbergen and DP0343398 with Zevenbergen and Lerman. This project explicitly addresses the pedagogic relay in both mathematics and culture.

Importantly we had to outline the role that Steve would play in this project. We argued that Steve's role in the project as being:

> ... the development of research instruments for data collection and for the analysis of classroom interactions. He has worked extensively with Bernstein's theoretical model and the construct of the pedagogic relay – this framing is critical to the theoretical developments of this project. He has worked on many projects that are focused on equity in mathematics education. He will be involved in data collection, analysis and report writing. Lerman has

worked in many low-SES schools in inner London and worked considerably with Bernsteinian frameworks to understand the role of pedagogy in the construction of social and cultural differences. He provides the key framings and guidance to this understanding that will be used to frame the project in novel ways that have not been undertaken in this context.

The grant was successful and funded from 2007 to 2011 at a cost of AU $250,000. Steve played a key role in the success of this application. While the intent had been for Steve to be involved in data collection, the prohibitive cost of travel meant that this was not possible.

Over the next few years, we would meet at conferences or before/after a conference to discuss the project and share findings/progress. It was difficult to gain a sense of the issues without being in the situation, particularly for people who have not been to remote Indigenous communities. In order to reduce the exorbitant costs of travelling to remote communities, we determined that the researchers based in the UK (Steve) and US (Jo) would not participate in data collection. We had also to be mindful of the cultural sensitivities about Indigenous education and reducing the number on non-Indigenous fly in-fly out researchers involved in the project.

The "Maths in the Kimberley" Project resulted in many publications that were subsequently compiled into a book and distributed freely to government, Catholic and Independent schools. It was also circulated to politicians – both state and federal – whose portfolios aligned with the foci of the study, as well as to various government departments involved with Indigenous education.

Theory Building

Steve's role in these projects was key in terms of building frameworks to understand the practices that were documented. His use of Vygotsky and Bernstein were important in coming to understand our work in ways that only Steve could create. His contribution to the projects enabled new understandings to emerge about ways in which the practices in ICTs mediated learning, particularly in terms of the use of the new emerging technologies of interactive whiteboards. This framing was novel. Similarly, his work with Bernstein and the micro analysis of classroom interactions enabled much deeper analysis of practice than was possible using a Bourdieuian lens on the observations of classrooms. These two frameworks – Vygotsky and Bernstein – have been central to Steve's work over a number of decades and were invaluable in our work on these projects.

The Humble Scholar and Friend

While a book such as this is about celebrating the successes of an outstanding academic, what is even more important to the many people who have had their lives and careers influenced by Steve is that he is an amazingly gracious and humble person. Steve is an outstanding scholar with a breadth of knowledge that he willingly shares with colleagues and friends. As evidenced by the work in this chapter where he has been successful in securing two ARC grants in the both streams of funding, Steve's contribution to the mathematics education community spans many nations – even to the Antipodes.

Steve has an amazing sense of humour and one that often takes you by surprise. I enjoy retelling the following story and while there are some elements that may be slightly embellished, it captures Steve's love of Australia. On one visit to Australia, he had to pass through customs. Steve was asked the usual questions about food stuffs and criminal records – Australians are a little cautious about what and who comes into the island continent. In Steve's impeccable BBC accent, he responded to the question regarding whether or not he had a criminal history with the comment "I didn't know I needed one anymore to get into this country." For those unfamiliar with Australian history, the nation was established as a penal colony to cater for the overflow of criminals in England so we are seen as a nation of convicts. While this story is somewhat embellished, it is one that those who know Steve, will appreciate. He has the capacity to draw on local knowledge (such as the penal history of Australia) and turn it into a moment of humour. This very human quality makes working with Steve over so many years such a true joy.

Steve's influence in these Australian collaborations is obvious and has helped to generate some amazing research. It has been a pleasure to have been able to find excuses to work with Steve on some exciting research. It has been an honour and pleasure to have worked with him on many projects, to have shared many meals and talks at conferences, to have worked on papers and editorial tasks, to have had his considered input into the projects outlined in this chapter, but mostly, I have enjoyed his friendship. He is, as we would say in Australia, a "great mate". The notion of a "mate" is an integral part of the Australian psyche and Steve is one of the best friends a person could have. So while this chapter and the book as a whole is a tribute to him of his contribution to the mathematics education research community, what I have valued most is his friendship. My life is so much richer to have had Steve as a mate, and to have been a part of my family. I look forward to the future where we will not only share intellectual ideas, but photographs and stories of our grandchildren.

References

Australian Research Council. (2011a). *Discovery projects funding rule for funding commencing in 2012*. Canberra: Australian Government.
Australian Research Council. (2011b). *Linkage projects: Funding rules for funding commencing in 2012*. Canberra: Australian Government.
Australian Research Council. (2012). *Linkage projects*. Retrieved October 10, 2012, from http://arc.gov.au/ncgp/lp/lp_default.htm
Bernstein, B. (1990). *The structuring of pedagogic discourse*. London: Routledge.
Bernstein, B. (1996). *Pedagogy, symbolic control and identity: Theory, research, critique*. London: Taylor & Francis.
Boaler, J. (2008). Promoting 'relational equity' and high mathematics achievement through an innovative mixed ability approach. *British Educational Research Journal, 34*(2), 167–194.
Boaler, J., & Staples, M. (2008). Creating mathematical futures through an equitable teaching approach: The case of Railside school. *Teachers College Record, 110*(3), 608–645.
Cooper, B., & Dunne, M. (1999). *Assessing children's mathematical knowledge: Social class, sex and problem solving*. London: Open University Press.
Education Queensland. (2000). *New basics: Theory into practice*. Brisbane: GoPrint.
Jorgensen, R., Sullivan, P., Grootenboer, P., Lerman, S., & Boaler, J. (Eds.). (2011). *Mathematics in the Kimberley*. Brisbane: Griffith Institute for Educational Research.
Lerman, S. (1988). *A social view of mathematics – Implications for mathematics education*. Paper presented at sixth international congress on mathematical education, Budapest.
Lerman, S., & Zevenbergen, R. (2006). Maths, ICT & pedagogy: An examination of equitable practice across diverse settings. In J. Novotna, H. Morova, M. Kratka, & N. Stehlikova (Eds.), *Mathematics in the centre: Proceedings of the 30th conference of the international group for the Psychology of Mathematics Education* (Vol. 4, pp. 49–56). Czech Republic: PME.
Lerman, S., & Zevenbergen, R. (2007). Interactive Whiteboards as mediating tools for teaching mathematics: Rhetoric or reality? In *Proceedings of the 31st conference of the international group for the psychology of mathematics education* (Vol. 3, pp. 169–176). Korea: PME.
Sullivan, P., Jorgensen, R., Boaler, J., & Lerman, S. (2013). Transposing reform pedagogy into new contexts: Complex instruction in remote Australia. *Mathematics Education Research Journal, 25*(1), 173–184.
Zevenbergen, R., & Lerman, S. (2001). Communicative competence in school mathematics: On being able to *do school* mathematics. In J. Bobis, B. Perry, & M. Mitchelmore (Eds.), *Numeracy and beyond: Proceedings of the 24th annual conference of the Mathematics Education Research Group of Australasia* (Vol. 2, pp. 571–578). Sydney: MERGA.
Zevenbergen, R., & Lerman S. (2005). Classroom interactions: Scaffolding students' Learning with ICTS. In M. Goos, C. Kanes & R. Brown (Eds.), *Proceedings of the 4th international conference of the Mathematics Education and Society Group*. Brisbane: Centre for Learning Research.
Zevenbergen, R., & Lerman, S. (2006). Maths, ICT and pedagogy: An examination of practice. In P. Grootenboer, R. Zevenbergen & M. Chinnapan (Eds.), *Proceedings of the 29th annual conference of the Mathematics Education Group of Australasia*. Canberra: MERGA.
Zevenbergen, R., & Lerman, S. (2007). Pedagogy and interactive whiteboards: Using an activity theory approach to understand tensions in practice. In K. Beswick (Ed.), *Proceedings of the 30th annual conference of the Mathematics Education Group of Australasia*. Hobart: MERGA.
Zevenbergen, R., & Lermans, S. (2008). Learning environments using interactive whiteboards: New learning spaces or reproduction of old technologies? *Mathematics Education Research Journal, 20*(1), 107–125.

Chapter 9
Researching the Role of the Teacher in Creating Socially Productive Classrooms that Facilitate Mathematics Learning

Peter Sullivan

Introduction

The explication by Steve Lerman of a social perspective on teaching and learning informed a classroom based research program including five funded projects. This chapter elaborates Lerman's perspective, illustrates how it informed each of the projects, and presents a specific lesson to exemplify the key elements of this social perspective. The chapter is a description of ways that aspects of Steve's 1998 plenary address at the conference of the *International Group for the Psychology of Mathematics Education*, titled "A moment in the zoom of a lens: Towards a discursive psychology of mathematics teaching and learning" (Lerman 1998) influenced a program of classroom based research. There are three themes in particular from that address that informed the research program described in the chapter: the search for a descriptive professional language; defining a role for the teacher; and a focus on the social context of the classroom.

In the first theme in that address, Lerman (1998) argued that "the concern of workers in the field is to find a language with which to describe the process of the acquisition of mathematics, and through which to draw inferences for what teachers might do to bring about that acquisition by as many students as possible" (p. 66). Within this quote there are three specific calls to action. The first call is associated with the search for a descriptive professional "language". In many other fields informed by empirical study, there is substantial attention to clarity and meaning of language and technical terms. In the field of mathematics education, there seems to be a startling reluctance by many researchers to build on definitions and terminology of others. The important attempts to introduce clear-planning and teaching terminology such as by Cobb and McClain (1999) on the social norms of the classroom, by Simon (1995) on the hypothetical learning trajectory and by

P. Sullivan (✉)
Faculty of Education, Monash University, Melbourne, VIC, Australia
e-mail: peter.sullivan@monash.edu

P. Gates, R. Jorgensen (Zevenbergen) (eds.), *Shifts in the Field of Mathematics Education*, DOI 10.1007/978-981-287-179-4_9

Brousseau (1997) on the didactical contract are important exceptions. While researchers can readily contrast comparable terms and consider subtlety in differences in usage, such nuances are often opaque to practitioners, contributing to what seems like a limited uptake of research ideas in classrooms. Indeed, the call by Steve Lerman for the development of descriptive language is probably as relevant now as it was then.

The second call is related to what "teachers might do" indicating that a goal of research in our field is to inform actions by teachers. While it is important that there are researchers exploring theoretical ideas, it is also critical that the focus of much of the research in mathematics education is to inform the practices of curriculum developers, teacher educators and teachers. Indeed such implications should be made explicit in reports of research.

The third call is around "acquisition by as many students as possible" which represents an explicit commitment to inclusiveness. The implication is that learning mathematics can create opportunities, and that the opportunities should be made available to the greatest number of students. There are clearly differences between the achievements of particular groups of students based on familiarity with the language of instruction, culture, socio-economic status, geography and gender. Even within each of these groups there is substantial diversity in achievement indicating that the search for pedagogies, that include as many students as possible, should be the focus of attention by education researchers.

Returning to the address, delivered at the height of the struggle in the field to discern the practical implications from the theory of constructivism, a second theme suggested by Steve Lerman was represented by the following statement: "the metaphor of students as passive recipients of a body of knowledge is terribly limited: so too is the metaphor of students as all-powerful constructors of their own knowledge, and indeed of their own identities" (p. 70). The former of these comments was addressing the all-too-common approach, especially in senior secondary and tertiary mathematics teaching, in which the fundamental planning process seems to be searching for clear explanations and ways of demonstrating mathematics concepts, following by repetitive practice. While this may be moderately successful with students who have elected mathematics study, it is hardly applicable for learners in earlier school years. It also results in a narrow set of experiences for successful learners. The latter metaphor addressed by Lerman was referring to the way that the radical version of constructivism (see, for example, Ernest 1994) was sometimes interpreted by classroom teachers as students working on potentially rich experiences without the benefit of expert teacher direction. Steve elaborated the Vygotskian opposition to pedagogies that seem to require students to "rediscover the development of mankind for themselves" (p. 69). Rather, Lerman argued that mathematics learning is centrally concerned with "the mediation of cultural tools and of metacognitive tools" (p. 69). For both of these, some explicit teacher guidance is needed.

In a third theme in the address, Steve emphasised "the centrality of the social relationships constituted and negotiated during classroom learning" (p. 70). Rather than attention to individual sense-making, he described an emerging process (at that time) in which the psychology of mathematics education is "focused on the way in which consciousness is constituted through discourse" (p. 67). In clarifying this social perspective, Steve elaborated the frequently cited the Zone of Proximal Development (ZPD) metaphor (Vygotsky 1978). Lerman argued that "the ZPD is created in the learning activity, which is a product of the task, the texts, the previous networks of experiences of the participants, the power relationships in the classroom, etc" (p. 71). Even though ZPD is sometimes used to describe teacher choice of an activity to allow students to step onto the next rung on a ladder of many miniscule steps of mathematics learning, Steve argued rather that ZPD is connected to creating classroom environments with conditions that are likely to facilitate student engagement in tasks that they have potential to complete. As he went on to argue "creating ZPD is more about mutual orientation of goals and desires than about the intended content of the interaction" (p. 72).

These three insights from that PME address – development of descriptive professional language, the role of the teacher in supporting student learning, and consideration of the social aspects of the classroom - were important in influencing the development of a program of research focusing on the effective use of tasks in classrooms. Basically it was assumed that the teacher has an active role in choice, adaptation and presentation of tasks, and the management of activity, and that the tasks should stimulate peer interactions as prompts to teaching and learning. The ways that these insights informed the individual projects within a program of research are elaborated in the following sections.

The Overcoming Barriers to Mathematics Learning Project

Supported by two successive grants from the Australian Research Council (ARC), together with Judy Mousley and Robyn Zevenbergen (now Jorgensen) I worked with teachers in their classrooms to describe approaches to teaching that were effective in supporting all students in their learning. One report of this research (Sullivan et al. 2003) described five elements of planning and teaching that they argued can be included in everyday routines of teachers. In our research we found that each of the five elements is manageable in classrooms, that teachers learn them readily, and that the elements can become part of the planning routines of teachers. The five elements are described in the following. The connections to Lerman's themes are considered subsequently.

Building a Communal Classroom Experience

In Sullivan et al. (2003) we argued that all students should have at least some core experiences that can form the basis of later discussions. The expectation is that teachers work with students to develop in them a sense of membership of the class as a whole. This notion is based on Wood (2002) who emphasised the way that "social interactions with others substantially contribute to children's opportunities for learning" (p. 61) and the interplay between children's developing cognition and the "unfolding structure that underlies mathematics" (p. 61). It was assumed that mathematical communications in classrooms that are intended to include all students can best occur if there is some communal experience. If some students in a class are excluded from common experiences and are unable to participate in discussion, this voids the possibility of them feeling affiliated with the class as a whole. Further, such experiences need not only create opportunities for social interaction but also promote thinking about mathematics.

Very much related to building this sense of communal experience is the need for the teacher to address diversity in mathematics awareness and attitudes. Such diversity can be a product of students' prior mathematical experiences, their familiarity with classroom processes (e.g., Delpit 1988), social, cultural and linguistic backgrounds (e.g., Zevenbergen 2000), the nature of their motivation (e.g. Middleton 1995), persistence and efficacy (e.g., Dweck 2000), and a range of other factors.

Some teaching approaches described elsewhere seem either to ignore the diversity of backgrounds and needs of students, or to address diversity in ways that exacerbate differences by, for example, having different goals for particular groups of students. In the Sullivan et al. approach, an expectation is that all students engage sufficiently in the goal or focus task to allow them to participate in a class review of the products of their work on that task so they participate in the social community that is the classroom (see also Askew et al. 1997).

Planning a Trajectory of Mathematical Tasks

Sullivan et al. (2003) argued that there are two considerations for the trajectory of tasks. The first is that there are benefits to inclusivity if at least some of the tasks are open-ended. It has been argued that open-ended tasks engage students in thinking about mathematics exploration, enhance motivation through increasing sense of control, and encourage students to investigate, make decisions, generalise, seek patterns and connections, communicate, discuss, and identify alternatives (Christiansen and Walther 1986; Middleton 1995; Sullivan 1999).

The second consideration relates to sequencing of tasks planned to offer the experiences necessary for students to complete the goal task. Simon (1995) described a hypothetical learning trajectory as one that:

> provides the teacher with a rationale for choosing a particular instructional design; thus, I (as a teacher) make my design decisions based on my best guess of how learning might proceed. This can be seen in the thinking and planning that preceded my instructional interventions … as well as the spontaneous decisions that I make in response to students' thinking. (pp. 135–136)

Simon noted that such a trajectory is made up of three components: the learning goal that determines the desired direction of teaching and learning, the activities to be undertaken by the teacher and students, and a hypothetical cognitive process, "a prediction of how the students' thinking and understanding will evolve in the context of the learning activities" (p. 136). These predictions are not dependent on students listening to a sequence of explanations but to engaging with a succession of problem-like tasks, based on recognition that learning and knowing is a product of activity that is "individual and personal, and ... based on previously constructed knowledge" (Ernest 1994, p. 2).

The intention is that earlier tasks in the sequence provide experiences that scaffold the student in the solution of later tasks, allowing them to engage in more sophisticated mathematics than would otherwise have been the case.

There are different ways to create sequences of tasks. One of these types of sequence is where the problem formulation remains constant but the numbers used increase the complexity of the task, say moving from small numbers to larger numbers. Another type of sequence is where the problem is progressively made more complex by the addition of supplementary steps or variables, such as in a network task where additional nodes are added. A third type of sequence may be where the concept itself becomes more complex, such as in a sequence of finding areas or progressively more complex shapes from rectangles, to composite shapes, to irregular shapes. The creation of such sequences is a key component of the planning model.

Enabling Prompts that Engage Students Experiencing Difficulty

A third element described by Sullivan et al. relates to supports offered to students who experience difficulty along the way, termed *enabling prompts*. It is common, indeed in places recommended, that teachers gather such students together and teach them (see, for example, Department of Education, Employment and Training 2001). Even worse is the practice of grouping students by teacher perception of their ability. It seems that the consensus is that this practice has the effect of reducing opportunities especially for students placed in the lower groups (Zevenbergen 2003). This can be partly due to self-fulfilling prophesy effects (e.g., Brophy 1983), and partly due to the effect of teacher self-efficacy which is the extent to which teachers believe they have the capacity to influence student performance (e.g. Tschannen-Moran et al. 1998). Brophy argued that, rather than grouping students by their achievement, teachers should: not worry too much about individual differences; keep expectations for individuals current by monitoring progress carefully; let progress rates rather than limits adopted in advance determine how far the class can go; prepare to give additional assistance where it is necessary; and challenge and stimulate students rather than protecting them from failure or embarrassment.

Students are more likely to feel fully part of the class if teachers offer prompts to allow those experiencing difficulty to engage in active experiences related to the initial goal task, rather than, for example, requiring such students to listen to additional explanations or assuming that they will pursue goals substantially different from those of the rest of the class. There are some generic types of prompts. For example, it nearly always helps to draw a diagram or model, to remove one of the constraints, to offer more choice, or to change the form of representation.

Tasks to Extend Thinking

A fourth element relates to anticipating that some students may complete the planned tasks quickly, and can be posed supplementary tasks that extend their thinking on that task. One of the characteristics of open-ended tasks is that they create opportunities for extension of mathematical thinking, since students can explore a range of options as well as considering forms of generalised response. In practice it is arguable that this is the most important and challenging of these planning steps. The premise is that the class progresses together through the lesson contributing to the sense of communal experience. Unless creative opportunities are provided for the students who have completed the tasks along the way then not only might they be bored, and so create difficulties for the teacher, but also they will not be using their time effectively. Note that this offers substantial advantages over the strategy of moving students who finish the work onto the next chapter of the text. Some strategies that teachers used to extend students' thinking included asking them to find all possible answers, to describe the possible answers generally, to create similar problems for other students, and to increase the complexity of the numbers involved or the number of problem steps.

Making Otherwise Hidden Pedagogies Explicit

The fifth element of the framework was related to being clearer about the pedagogies of mathematics teaching. This was informed by the work of Bernstein (1996) who described pedagogies that are hidden from some students. Bernstein argued that, through different methods of teaching, students receive different messages about the overt and the hidden curriculum of schools. He suggested that some students are able to make sense of this "invisible" pedagogy more effectively than others, due to their familiarity with the embedded socio-cultural norms, and hence those students have more chance of success.

As suggested by Delpit (1988), Zevenbergen (1998), and Dweck (2000), it may be possible to moderate the effect of the hidden curriculum by explicit attention to aspects of pedagogies associated with such teaching. Sullivan et al. (2002) listed a range of particular strategies that teachers could use to make implicit pedagogies more explicit and so address aspects of possible disadvantage of particular groups. It seems that teachers are able to make explicit at least some of the key pedagogies associated with such teaching, and that students respond to this explicitness in the direction intended (Sullivan et al. 2003). Zevenbergen et al. (2004) describe three strategies used by one teacher in a school with a high proportion of Indigenous students to make particular pedagogies explicit, either by stating the issue directly or through modelling of a socio-cultural process.

In summary, the three themes from Steve Lerman were foundational in the identification and description of the five elements of this pedagogical approach. There are clearly articulated roles for teachers and students that emphasise teacher/student interactions and student decision-making, there is a explicit intent to include all students in the mathematics learning, key concepts such as classroom social norms and learning trajectory were utilised, and the idea of enabling and extending prompts that drew on Steve's interpretation of ZPD were proposed.

This approach also emphasises the creation of a classroom community and emphasises the social aspect of learning. Various reports of this project (e.g., Sullivan et al. 2009a) not only provide evidence of the feasibility and impact of the approach, but also provide classroom examples that exemplify enacting the planning model.

Students' Responses to Different Types of Mathematics Tasks

Subsequent to the Overcoming Barriers project, I worked with Doug Clarke and Barbara Clarke on the *Task Types in Mathematics Learning (TTML)* project which was a 3 year Australian Research Council funded research partnership between the Victorian Department of Education and Early Childhood Development, the Catholic Education Office (Melbourne), Monash University and Australian Catholic University. The TTML project was strongly influenced by the findings of the earlier project, and also sought to build not only on Lerman's consideration of the role of the teacher, but also on what ZPD might mean in the broader context of converting potentially rich tasks to learning experiences (for a full report on this project, see Sullivan et al. 2013).

The project worked with clusters of middle-years' teachers of mathematics (Grades 5–8) with each cluster typically involving a secondary school and three or four primary schools. These three clusters represented a spread of socio-economic student backgrounds and included schools in both government and Catholic systems. The clusters examined the processes of identifying potentially rich tasks, creating lessons and sequences of lessons from those tasks, and exploring the pedagogies that are associated with the active engagement and support of all students. The notions of enabling prompts and extending prompts, which built directly on Steve's explication of ZPD, were prominent.

Of particular interest was the process of developing and communicating classroom social norms. To explore this further, we sought responses from students to some surveys to gain insights into the types of tasks they value, the nature of the lessons they prefer, and their aspirations in relation to classrooms. The data from the surveys is reported in full in Sullivan et al. (2013), but the following is an attempt to summarise some relevant findings.

The first finding was that while there was not much difference overall in the students' reported confidence and satisfaction over the years 5, 6, 7 and 8 (ages 10–13), even though they move from primary to secondary schools after year 6. At each of these middle-years' levels there is a range of student satisfaction and confidence. There were also substantial and significant differences between classes suggesting that teachers have a major impact of the satisfaction and confidence of their students. In terms of creating a classroom that has socially inclusive norms, teachers should be aware of the views of each of their students. It seems it would be productive for teacher educators to support teachers with strategies for finding out students' levels of satisfaction and confidence, and also for making suggestions on moving students in the direction of being more satisfied and more confident.

A second finding related to different types of tasks ranging from those that focused on mathematical principles to those that were based on realistic contexts to those which were investigative or open-ended. The survey suggested that each of the task types was liked most by some students, and likewise each of the types of task was rated as the one from which they could most learn. The implication is that, in creating an inclusive approach, it is essential teachers use a range of types of tasks in their teaching. This may be particularly relevant to teachers at the secondary schools who seem to use texts with mainly similar types of tasks. This finding also suggests that students need support to gain benefits from tasks that they do not like or do not feel that they can learn from. Teachers may well benefit by making students aware of the purpose of tasks and what it is that teachers are hoping students will learn from them. In terms of the third of Lerman's themes, creating a socially inclusive classroom means that the purpose of pedagogies, the social norms and expectations for effort need to be explicit.

A third finding was drawn from free format essays written by students on the type of lessons they prefer. The essay responses were categorised and quantified. One set of categories, that represented around half of the comments overall, related to lessons that have engaging pedagogies. There was also around half of all the comments that mentioned specific aspects of the mathematical content. Teachers need to find ways to focus on content using engaging and with the purpose and connection of each of these articulated explicitly to the students.

A fourth finding, also from the free format essays, was that the method of grouping is important for students, and nearly all students mentioned grouping in their responses. Most students prefer to work in groups or pairs in their mathematics classes, but around 20 % of the close to 1,000 students indicated unprompted that they prefer to work alone. For similar reasons as with the previous finding, teachers need to explain the method of grouping they are using, and the purpose of those groupings, to the students. It is also helpful to teach specific social skills such as listening to others.

A fifth finding is that the ways that teachers interact with students is important for them. The free format essays indicated that there is a variety in student preferences, so it would be useful for teachers to find out the types of interactions that individual students find helpful.

A sixth finding is that there was tension between students' liking of an approach and the extent to which they felt it helped them learn. This also suggests that teachers need to be explicit about their intentions related to students' satisfaction and confidence, posing a variety of types of tasks, the focus of lessons whether on content or engagement, the method of grouping, and modes of support and interaction.

Connected to the third of Steve's themes, these results elaborate the ways that social considerations inform and influence the classroom community and the learning opportunities of the students. The emphasis in the unprompted responses from the students is on relationships and connections. It seems that descriptive language may well assist students in aligning their efforts with those of the teacher, that teacher guidance is important, and that facilitation of the social dimension of learning is a key challenge for teachers.

The Connection Between Challenge and ZPD

A further ARC-funded project that builds not only on the three key themes from Steve Lerman described earlier but also the above two projects is *Encouraging Persistence Maintaining Challenge* (EPMC) in which I am working with Jill Cheeseman, Doug Clarke, Angela Mornane, Jim Middleton, and Anne Roche. The rationale of the project is described as:

> ...founded on a belief that while it is possible for everyone to learn mathematics, it takes concentration and effort over an extended period of time to build the connections between mathematical ideas, and to be able to transfer learning to practical contexts and new topics. The type of effortful actions that are associated with learning mathematics include connecting, representing, identifying, describing, interpreting, sorting, applying, designing, planning, checking, imagining, explaining, justifying, comparing, contrasting, inferring, deducing and proving. ...All of these actions or activities require effort to initiate and modulate behaviour, and persistence to maintain and direct behaviour towards successful ends.

The project team describes *persistence* as student actions that include concentrating, applying themselves, believing that they can succeed, and maintaining effort to learn. We call tasks that are likely to foster such actions *challenging*.

The project is examining the rationale and processes for encouraging students to persist when working on challenging mathematical tasks, the nature of tasks that provide the appropriate type of challenge, and the actions that teachers can take to support students engaged in challenging tasks.

Primarily the project is exploring student persistence, the opportunities and constraints it creates, and ways of fostering it. We are assuming that this involves teachers posing challenging tasks, encouraging students to take risks in their learning and to work with other students, to allow them make their own decisions on ways of solving problems and to create opportunities for students to explain their reasoning and justify their thinking. We are examining teacher actions that might encourage students to persist.

An aspect of this project that focuses on the student contribution to the social norms of the classroom related to the potential constraining effect that students can sometimes have on other students and the need to address this. One useful model informing our approach to addressing this was proposed by Dweck (2000) who explained that students who have a performance orientation, meaning they seek social affirmation as the goal of their effort rather than understanding of the content, avoid risk taking and challenging tasks due to fear of failure. Sullivan et al. (2009b) described a powerful classroom culture in which students fear censure from peers if they appear to persist, and so avoid the appearance of trying hard.

The first phase of the project involves design projects that included two iterations of observations of a group of teachers who posed challenging mathematical tasks and took actions to encourage students to persist on those tasks. In reporting on the analysis of those classroom observations we noted:

> The teachers were all willing to try the tasks suggested which may have involved departures from their usual practice. While the lessons may not have been perfect, they each created important opportunities for student learning. . . . The task suggestions presented to teachers were welcomed by the teachers and allowed them to adapt the task for themselves. The teachers generally maintained the challenge of the task, and encouraged the students to persist. The teachers adapted the suggested tasks to suit their preferred lesson structure, and it was noted that managing student engagement seems to be important. The lesson review phase is clearly a critical feature of the use of challenging tasks. (p. 44)

Again the three themes from Lerman are evident. The development of a descriptive language around persistence and challenge is important, as is communicating the intent of those terms to both teachers and students. It is clear that the posing of challenging tasks involves neither explicit instruction nor expectations that students create the relevant knowledge for themselves. It is evident that the social context in which challenging tasks are posed both offers opportunities and creates constraints.

The Maths in the Kimberley Project

I was invited to join the *Maths in the Kimberley* project, led by Robyn Jorgensen (see Jorgensen 2009), as a result of this ongoing work on tasks. The project explored three key pedagogical approaches to mathematics teaching and learning:

- ensuring that the focus of learning is both mathematically rigorous and culturally appropriate;
- not only building a sense of community learning and but also accommodating variations to suit particular learners' needs and experiences;
- recognising that teachers' knowledge, beliefs, and expectations have an impact on learners and the key element in any attempt to improve learning is the teacher.

The project examined these issues, both individually and together, in schools in the Kimberley region of Western Australia in which, arguably, the challenge of teaching mathematics is profound due to the multiplicity of variables that impact on the quality and outcomes of learning opportunities. The goal of the research was to investigate the challenges and opportunities afforded by our pedagogical approach.

The project found that being clear about the goals of teaching is helpful both to the teacher and to the students. It appears that "building on what the students know" is an effective strategy. Students are both willing and able to engage with rich tasks that required decision making by them and which allowed the construction of mathematical ideas.

In Conclusion: What Might a Lesson Based on These Ideas Looks Like? Rather than presenting a summary of the above arguments, the following is an attempt to exemplify what the three themes from Steve Lerman's address might look like in classrooms. Indeed, in the 1998 address, Steve offered some insights into what teaching based on the three themes might look like. He wrote:

> When a teacher offers an activity in a classroom, say to share 2 oranges between 3 children, the different answers offered by the children arise from their previous experiences, what has been called the zone of actual development, and potentially pull the others including perhaps the teacher, into their zones of proximal development. (p. 77)

In other words, the teacher poses a task about which there is some ambiguity, supports student while they engage with the task, and then manages a discussion to which all students have the opportunity to contribute. To elaborate this, and to synthesise the ideas from the projects described above, the following is a description of a lesson that I observed when in Japan.

To establish a context or rationale for the learning, the teacher led a discussion about *tatami* mats (which are traditional rice-straw mats 90 cm

(continued)

by 180 cm used commonly and sometimes used to describe floor size (area) of rooms or large buildings).

Next the teacher posed the central problem for the lesson, which was to work out how many squares in the following diagram are shaded. The intent of the teacher was to introduce the notion that the number of squares in an array can be calculated by multiplying the number of rows by the number of columns. The teacher had presented students with a worksheet on which there were TWO examples of this diagram,

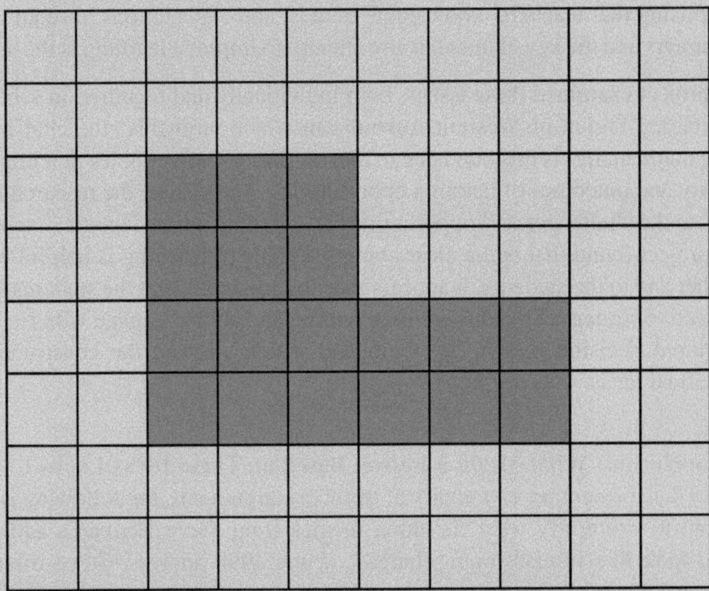

This had the effect of emphasising to the students that it was not the answer that is important but the method of solution that they would devise. Rather than reproducing the teacher's method, the students could explore their own solution, and so have something to contribute to a classroom discussion later.

While the students were working, the teacher moved around the class identifying students who would contribute to a whole class discussion of their strategies.

The selected students were then given an A3 size version of the diagram, on which they were invited to record their solution. The students then displayed their A3 size versions of their solutions, and wrote the calculation they used. The following photograph records two of the solutions, one of which involved breaking the shape into two rectangles, and in the other the student moved one part to create a new rectangle.

(continued)

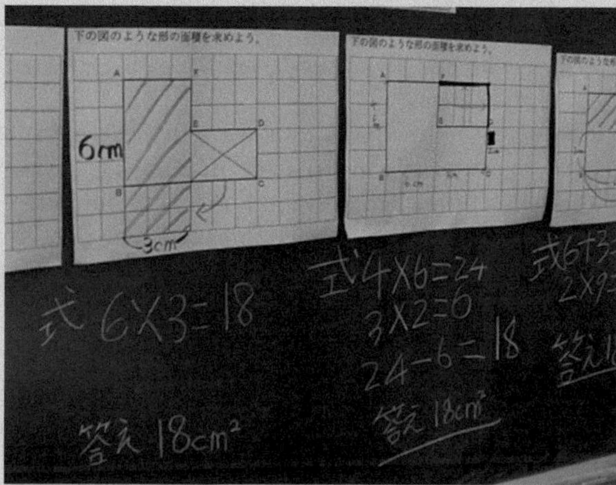

Eventually the board looked like the following with seven different solutions displayed and explained. The teacher had assisted each of the students to explain their strategy, and then spent some time drawing the key themes together.

There are some interesting characteristics of this lesson:

- there was an obvious focus on student generated strategies;
- in the introduction the teacher connected the content to a realistic context, the clarified the task, but did not predetermine the methods that the students would use;
- students were thoughtfully chosen to ensure that a range of strategies were presented and discussed;

(continued)

- the task is not complex pedagogically or organisationally, and only requires a willingness to let the students' thinking emerge;
- the task was accessible to all students, and the pedagogies supported the engagement of all students;
- the task was thoughtfully chosen to allow a diversity of strategies and representations, and also to allow students to experience important mathematical ideas such as area conservation (that is useful in the process of calculating the area of parallelograms), breaking a composite shape into parts (that can inform the calculation of the area of trapezia), and subtracting areas (that is used in calculating the area of paths around shapes).

Returning to the first of Steve's three themes, this lesson illustrates the call for the development of descriptive language. Interestingly, there are Japanese words that describe various aspects of this lesson. As described by Inoue (2010), *Hatsumon* refers to the initial problem; *Kizuki* describes what the teacher wants the students to learn; *Kikanjyuski* is the individual or group work on the problem; *Kikanshido* describes the teacher thoughtfully walking around the desks; *Neriage* is the carefully managed whole class discussion seeking the students' insights; and *Matome* is the teacher summary of the key ideas. No doubt this descriptive language is at the basis of the Japanese approach to lesson planning.

In terms of the perspective on how students come to learn mathematics, the content and methods were neither directed by the teacher nor did they involve undirected student activity. In fact, the task on which the students worked was sensitively chosen to facilitate student opportunity to create strategies, and the lesson was structured to facilitate the student contribution to the class learning.

It was clear that the class was operating as a community, and the social interactions at the time of lesson review, especially in the explanation of strategies and the opportunity for questioning of the presented strategy. The selected students clearly had expected this opportunity to explain their thinking, and the other students listened respectfully to the presenters, assuming presumably that this was an opportunity for them to learn.

This lesson illustrates that not only are the approaches recommended by Steve Lerman possible, they can result in meaningful student learning.

References

Askew, M., Brown, M., Rhodes, V., Johnson, D., & Wiliam, D. (1997). *Effective teachers of numeracy, final report*. London: King's College.

Bernstein, B. (1996). *Pedagogy, symbolic control, and identity: Theory, research, critique*. London: Taylor & Francis.

Brophy, J. E. (1983). Research on the self-fulfilling prophecy and teacher expectations. *Journal of Educational Psychology, 75*(5), 631–661.
Brousseau, G. (1997). *Theory of didactical situations in mathematics 1970–1990* (M. Cooper, N. Balacheff, R. Sutherland, & V. Warfield, Edited and Trans.). Dordrecht: Kluwer Academic Publishers. (French version (1998). Théorie des situations didactiques. [Textes rassemblés et préparés par N. Balacheff, M. Cooper, R. Sutherland, & V. Warfield]. Grenoble: La penséesauvage).
Christiansen, B., & Walther, G. (1986). Task and activity. In B. Christiansen, A. G. Howson, & M. Otte (Eds.), *Perspectives on mathematics education* (pp. 243–307). Dordrecht: Reidel.
Cobb, P., & McClain, K. (1999). Supporting teachers' learning in social and institutional contexts. In Fou-Lai Lin (Ed.), *Proceedings of the 1999 international conference on mathematics teacher education* (pp. 7–77). Taipei: National Taiwan Normal University.
Delpit, L. (1988). The silenced dialogue: Power and pedagogy in educating other people's children. *Harvard Educational Review, 58*(3), 280–298.
Department of Education, Employment and Training. (2001). *Early years numeracy strategy.* http://www.sofweb.vic.edu.au/eys/num/index.htm
Dweck, C. S. (2000). *Self theories: Their role in motivation, personality, and development.* Philadelphia: Psychology Press.
Ernest, P. (1994). Varieties of constructivism: Their metaphors, epistemologies and pedagogical implications. *Hiroshima Journal of Mathematics Education, 2*, 1–14.
Inoue, N. (2010). Zen and the art of neriage: Facilitating consensus building in mathematics inquiry lessons through lesson study. *Journal of Mathematics Teacher Education,* Retrieved online from http://www.springerlink.com/content/g33022h2k1384461/
Jorgensen, R. (2009). Cooperative learning environments. In R. Hunter, B. Bicknell, & T. Burgess (Eds)., *Crossing divides (Proceedings of the 32nd conference of the Mathematics Education Research Group of Australasia*, Vol. 1, pp. 700–703). Sydney: MERGA.
Lerman, S. (1998). A moment in the zoom of a lens: Towards a discursive psychology of mathematics teaching and learning. In A. Olivier & K. Newstead (Eds.), *Proceedings of the 22nd conference of the International Group for the Psychology of Mathematics Education* (Vol. 1, pp. 66–81). Stellenbosch.
Middleton, J. A. (1995). A study of intrinsic motivation in the mathematics classroom: A personal construct approach. *Journal for Research in Mathematics Education, 26*(3), 254–279.
Simon, M. (1995). Reconstructing mathematics pedagogy from a constructivist perspective. *Journal for Research in Mathematics Education, 26*, 114–145.
Sullivan, P. (1999). Seeking a rationale for particular classroom tasks and activities. In J. M. Truran & K. N. Truran (Eds.), *Making the difference. Proceedings of the 21st conference of the Mathematics Educational Research Group of Australasia* (pp. 15–29). Adelaide.
Sullivan, P., Zevenbergen, R., & Mousley, J. (2002). Contexts in mathematics teaching: Snakes or ladders? In B. Barton, K. C. Irwin, M. Pfannkuch, & M. Thomas (Eds.), *Mathematics education in the South Pacific* (pp. 649–656). Auckland: Mathematics Education Research Group of Australasia.
Sullivan, P., Mousley, J., Zevenbergen, R., & Turner Harrison, R. (2003). Being explicit about aspects of mathematics pedagogy. In N. A. Pateman, B. J. Dougherty, & J. T. Zilliox (Eds.), *Proceedings of the 2003 joint meeting of PME and PMENA* (pp. 267–275). University of Hawai'i, PME.
Sullivan, P., Mousley, J., & Jorgensen, R. (2009a). Tasks and pedagogies that facilitate mathematical problem solving. In B. Kaur & Association of Mathematics Educators (Eds.), *Mathematical problem solving* (pp. 17–42). Singapore/Hackensack: UK World Scientific Publishing.
Sullivan, P., Prain, V., Campbell, C., Deed, C., Drane, S., Faulkner, M., McDonough, A., Mornane, A., & Smith, C. (2009b). Trying in the middle years: Students' perceptions of their aspirations and influences on their efforts. *Australian Journal of Education, 5*(2), 176–191.

Sullivan, P., Clarke, D., & Clarke, B. (2013). *Teaching with tasks for effective mathematics learning*. New York: Springer.

Tschannen-Moran, M., Hoy, A., & Hoy, W. (1998). Teaching efficacy: Its meaning and measure. *Review of Educational Research, 68*(2), 202–248.

Vygotsky, L. S. (1978). *Mind in society: The development of higher psychological processes* (M. Cole, V. John-Steiner, S. Scribner, & E. Souberman Eds. and Trans.). Cambridge, MA: Harvard University Press.

Wood, T. (2002). What does it mean to teach mathematics differently? In B. Barton, K. C. Irwin, M. Pfannkuch, & M. Thomas (Eds.), *Mathematics education in the South Pacific* (pp. 61–71). Auckland: Mathematics Education Research Group of Australasia.

Zevenbergen, R. (1998). Language, mathematics and social disadvantage: A Bourdieuian analysis of cultural capital in mathematics education. In C. Kanes, M. Goos, & E. Warren (Eds.), *Teaching mathematics in new times* (pp. 716–722). Gold Coast: MERGA.

Zevenbergen, R. (2000). "Cracking the code" of mathematics: School success as a function of linguistic, social and cultural background. In J. Boaler (Ed.), *Multiple perspectives on mathematics teaching and learning* (pp. 201–223). New York: JAI/Ablex.

Zevenbergen, P. (2003). Ability grouping in mathematics classrooms: A Bourdieuian analysis. *For the Learning of Mathematics, 23*(3), 5–10.

Zevenbergen, R., Mousley, J., & Sullivan, P. (2004). Disrupting pedagogic relay in mathematics classrooms: Using open-ended Tasks with Indigenous students. *International Journal of Inclusive Education, 8*(4), 391–405.

Part V
Steve's Theoretical Contributions

Chapter 10
Turning Mathematical Knowledge for Teaching Social

Jill Adler

> *We might suggest that the field [of mathematics education research] exhibits a weak grammar, in that we can see a proliferation of new specialised languages, creating new positions within the field.*
>
> (Lerman et al. 2002, p. 37)

> *... [the] privileged position [of mathematics as a field of knowledge] can be seen to place mathematics education in great danger as the research community feels itself free to pursue "internal" issues of teaching and learning mathematics whilst policy makers put pressure on teachers to perform according to their own pedagogical and curricular demands ...*
>
> (Lerman 2014, p. 13)

Introduction

I select the above two quotations from Steve Lerman's work in mathematics education research as they structure and illuminate the two inter-related problems I pursue in this chapter. Furthermore, as with other chapters in this book, these quotes signal some of the contribution of Steve's research to the development of mathematics education research, and its critique. Signalled first for this chapter is a question about the research on 'mathematical knowledge for teaching' as a subdomain in the field of mathematics education, and so its grammar, specialised language, and the new positions created. Hence, the questions I pursue here are:

- What kind of knowledge is mathematical knowledge for teaching?
- Why does this knowledge matter?
- What new position(s) are opened?
- How do these feature in the problem of the 'internal' nature of research in mathematics education, and so too research on mathematical knowledge for teaching?

J. Adler (✉)
School of Education, University of the Witwatersrand, Johannesburg, South Africa
e-mail: jill.adler@wits.ac.za

P. Gates, R. Jorgensen (Zevenbergen) (eds.), *Shifts in the Field of Mathematics Education*, DOI 10.1007/978-981-287-179-4_10

I will develop and then reflect on two inter-related claims in this chapter. The first is that the sets of concepts that have emerged through research on mathematics knowledge for teaching (MKT), while relatively recent, have nevertheless proliferated. This is not surprising given that as part of educational knowledge, it is part of a horizontal knowledge structure with a relatively weak grammar (Bernstein 1999). The second is that a key 'new' position producing and produced by this knowledge development is that of *mathematics-teacher-educator-researcher* working simultaneously as knowledge producer *and* recontextualiser in the university. A number of questions, about research and practice emerge from the grammar of MKT and the dual, perhaps ambiguous positioning of its agents. This chapter thus offers a story about mathematical knowledge for teaching framed by Steve Lerman's contributions to the field, and the possibilities evoked for further work.

Mathematical Knowledge for Teaching – A Horizontal Knowledge Structure

I have already stated that as part of educational knowledge, MKT has a *weak grammar*, and concepts related to this notion have proliferated. This claim follows Bernstein's analysis of disciplinary discourses and knowledge structures (Bernstein 1999, 2000), an analysis that informed the study of the development of mathematics education research as a field (Lerman et al. 2002).

Briefly, Bernstein (2000) offers a set of theoretical resources for interrogating the production of knowledge. He distinguishes in the first instance between two major discourses within which knowledge circulates, grows and changes: vertical and horizontal. A similar distinction is made by many others (e.g. Vygotsky's concepts of the scientific and the everyday). Horizontal discourse "entails a set of strategies which are local, segmentally organised, context specific and dependent …", and vertical discourse is "a coherent, explicit and systematically organised structure" (op cit, p. 157). Bernstein then goes on to disaggregate vertical discourses, and the different modalities of knowledge realised within vertical discourses. Hierarchical Knowledge Structures, for example Physics, which are geared towards "greater and greater integrating propositions, operating at more and more abstract levels", and Horizontal Knowledge Structures, found within the Humanities and Social Sciences, which consist of a "series of specialised languages with specialised modes of interrogation and criteria for construction and circulation of texts". Within Hierarchical Knowledge Structures there is an integration of language, and ever increasing abstraction; development of a Horizontal Knowledge Structure, in contrast, entails the production of new languages.

A further distinction is then made within Horizontal Knowledge Structures, between disciplines like Economics and Linguistics on the one hand, where structures have a relatively 'strong' grammar; and others, like Sociology, a relatively 'weak' grammar. Education, in turn, forms a region, in Bernstein's terms, as it recruits languages from the Social Sciences, and as Lerman et al. (2002) show, the development of mathematics education research has drawn from an increasing array of languages within the Social Sciences. Education has a particularly weak grammar. Recognition of what is and is not the language of scholarship and knowledge development in education is contested and far less clear than mathematics itself, or physics, or economics. Moreover, what counts as legitimate educational knowledge is not only different across languages within education, but also ambiguous, and open to interpretation and so contestation. It is in this terrain that Bernstein himself as a sociologist of education worked to build a language of description for pedagogic discourse, so as to strengthen what Maton and Muller (2007) have called the verticality and grammaticality of this relay. As others argue (e.g. Lemke 1993), it is through stronger grammars which enable unambiguous descriptions that disciplines grow. Growth of educational knowledge too, will thus benefit from greater verticality and grammaticality.

In Bernstein's terms then, MKT is part of region (Education), which in turn draws on multiple Horizontal Knowledge Structures (e.g. psychology, sociology), and through this MKT too is likely to be constituted by a proliferation of concepts and a weak grammar.

Multiple Frameworks of MKT as Knowledge-in-Use

My concern in this chapter is mathematical knowledge for teaching (MKT), and so the questions of interest are, what kind of knowledge is this, and why does it matter? The current focus on mathematics teachers' knowledge in the field is evident in special issues and a range of research papers across key journals. Two recent issues of the journal *Zentralblatt für Didaktik der Mathematik* (now: ZDM – The International Journal on Mathematics Education) have focused on teacher expertise (Volume 43, Issue 6–7, November 2011) and measuring MKT across contexts (Volume 44, Issue 3, 2012). A paper on knowledge for teaching algebra has just been published in the *Journal for Research in Mathematics Education*, and while one would expect the *Journal of Mathematics Teacher Education* with its focus on teacher education to include papers on teachers' knowledge, it is interesting to see a focus on teachers' knowledge, practice, and identity in Volume 16, Issue 6, 2011; and teacher knowledge as fundamental to effective teaching practice in Volume 15, Issue 3, 2012.

With this elaboration in the field, has come a proliferation of concepts and frameworks. It is useful to distinguish two lines of research. The first, following or developing from Shulman (1986, 1987) has focused on describing the specificity of MKT, with descriptions emerging from empirical research on knowledge-in-use in the practices of mathematics teaching. The underlying assumption here is that it is from studies of mathematics classroom practice, that is, of teachers teaching mathematics in school, and other records of mathematics teaching, that one 'finds' mathematical knowledge for teaching. We can include here:

- the extensive research work on MKT by Deborah Ball and her colleagues in Michigan elaborating on MKT as including distinctions within Shulman's notions of subject matter and pedagogic content knowledge (Ball et al. 2008);
- the study of Liping Ma (1999) and her elaboration of 'deep' subject knowledge' as PUFM – profound understanding of mathematics – and its four further properties: connectedness; multiple perspectives; basic ideas; longitudinal coherence (p. 122);
- the elaboration of 'mathematics for teaching' by Davis (2011); and
- the study of Rowland et al. (2005) and the development of the 'knowledge quartet' as rubric for researching and reflecting on practice. Acts of mathematics teaching that foreground content knowledge in use for Rowland et al. include drawing on 'transformation', 'connections', 'contingency' and 'foundational knowledge'.

Each of these four studies, while acknowledging and referring to each other's work, provide their own conceptual frame, designed for or through their particular study and question – and so the proliferation of language.

Measurement Research on MKT – Is This Strengthening the Grammar?

A comprehensive review of research on assessing MKT in the US, focused on "what knowledge matters and what evidence counts", traces the development of methods for describing and measuring professionally situated mathematical knowledge in the United States (Hill et al. 2007a). As elaborated elsewhere (Adler and Patahuddin 2012), Hill et al. locate their recent measures work done in the Learning Mathematics for Teaching (LMT) project, in the context of the qualitative research of the 1980s and 1990s, building from its successful but small scale developments to enable large scale, reliable and valid ways of assessing professionally situated knowledge. The results of the LMT research have been widely published and include reflection on how, building from Shulman's (1986) initial work, the development of measures simultaneously produced an elaboration of the construct MKT and its component parts. As they developed measures, they were able to distinguish and describe Subject Matter Knowledge (SMK) and Pedagogic Content Knowledge (PCK), and categories of knowledge within each of these domains. Common Content Knowledge (CCK – mathematics that might be used across a range of practices) was delineated from Specialised Content Knowledge (SCK – mathematics used specifically in carrying out tasks of teaching) (Ball et al. 2008). Within PCK, where knowledge of mathematics is intertwined with knowledge of teaching and learning, they distinguish Knowledge of Content and Students (KCS – e.g. knowledge about typical errors learners make, or misconceptions they might hold), from Knowledge of Content and Teaching (KCT – e.g. knowledge of particular tasks that could be used to introduce a topic). In addition to describing their MKT constructs and exemplifying measures of these, they have reported on positive correlations they found in their study of the relationship between measures of teachers' MKT, the quality of their mathematics teaching and their learners' performance (Hill et al. 2005, 2008).

In their concern for construct validation, the LMT project has subjected its work to extensive critique. A whole issue of *Measurement* (Vol. 5, No 2–3, 2007) addressed this purpose. Difficulties entailed in measures work are critiqued within the LMT project itself, particularly PCK items aimed at KCS (Hill 2008; Hill et al. 2007b). The strength of the construct of PCK, in their terms, depends on how well it can be distinguished from knowledge of the mathematical content itself. LMT validity tests, including clinical interviews on these items, failed to separate KCS from related measures of content knowledge. Scores on KCS items correlated highly with CCK scores. Hill et al. (2007a, b) and Hill (2008) describe additional insights from their cognitive interviews on PCK- KCS items that showed that teachers also used mathematical reasoning, and test-taking skills, to decide on the correct answer. Hill et al. (op cit) conclude that "this domain [PCK] remains underconceptualised and understudied" (p. 395), despite wide agreement in the field that this kind of knowledge matters. Their reflection on their detailed PCK work highlights difficulties in operationalizing strong metaphorical notions like PCK. As a field, we continue to use such notions as if they were clear, and empirical recognition relatively straight forward.

Construct delineation and validation is a strong feature of quantitative research, and central to the work of (Krauss et al. 2008) in their large scale study of secondary mathematics teachers' professional knowledge and its relationship to learner performance. Based in Germany, their measure development and use in the COACTIV project, like Hill et al., worked from the assumption that professional knowledge is situated, specialised, and thus requires assessments that are not synonymous with tests at particular levels of institutionalised mathematics (be this school or university). Indeed, for Krauss et al., secondary teachers' SMK (what they call Content Knowledge – or CK) sits in a space between school mathematics and tertiary mathematics (p. 876), and is clearly bounded from their interpretation of PCK. They conducted CK and PCK tests on different groups selected with respect to professional knowledge (i.e. mathematical knowledge in and for teaching): and results confirmed their professional knowledge hypothesis – experienced teachers irrespective of their teacher education route showed high PCK scores. At the same time, however, mathematics major students performed unexpectedly well on PCK items. Krauss et al. (op cit, p. 885) explore this interesting outcome in their study – how it was that mathematics major students, who had no teaching training or experience, were relatively strong on their PCK items.

Of interest in this chapter is the analysis of the diverse ways in which professional knowledge constructs have been operationalized in the field. Krauss et al., for example, exemplify a PCK task item that asks: "How does the surface area of a square change when the side length is tripled? Show your reasoning. Please note down as many different ways of solving this problem as possible". The sample response given includes both an algebraic and geometric representations (p. 889).

In Ball et al.'s terms, this response does not require specific or local knowledge of students, nor of curricula, or particular teaching tasks, and hence, in their terms would be SCK, and distinct from PCK. We concluded that:

> "knowledge of multiple representations shifts between PCK and SMK across these two studies … [and that the MKT]" construct and its components are differently operationalised in different studies, a point made by Hill et al., (2007a, b) and noted as a shortcoming in this research. (Adler & Patahuddin, op cit)

Thus, even in studies where operationalization for measurement purposes is critical, elements of a weak grammar (multiple meanings for the same concept) in our field are thus evident.

From Knowledge in Use to Knowledge Produced

In contrast to the studies of MKT with mathematics teaching practices as the empirical field, our study of MKT in the QUANTUM project (cf. Adler and Davis 2006, 2011; Parker and Adler 2014) was undertaken in the field of mathematics teacher education. Our interest was in describing what and how MKT is constituted in and across ranging contexts of mathematics teacher education, and so how such a notion is taking shape in mathematics teacher education practice. We have examined pedagogic discourse as this unfolds in pedagogic practice across various courses so as to describe what is legitimated as mathematics for teaching (MfT) and how this occurs. In developing our methodology, we built from an assumption that in mathematics teacher education, both 'mathematics' and 'teaching' are objects of learning. Depending on the focus of activity, however, either mathematics, or teaching, will be the primary object, with the other likely to be present yet back-grounded. We represented this simultaneous privileging and back-grounding as Mt, or Tm, where the capitalisation marks the privileging, and simultaneously weakens the boundary between SMK and PCK. This co-constitution has effects on what and how mathematics and/or teaching mathematics and so MKT is made available to learn in mathematics teacher education practice.

This work developed at the same time as the knowledge-in-use research discussed above, and attempted to connect with and contribute to its development. In our early work, (Adler and Davis 2006) we referred to MKT as simply 'mathematics for teaching' and described it as a "new and fledgling discourse". A particular concept that we worked to develop was Ball et al. (2004) notion of "unpacking". Ball et al. used the notion of unpacking to illuminate some of the specialised mathematical work of teaching that marks it out as distinct from the mathematical work of mathematicians. While the hallmark of development of mathematics, and so the work of mathematicians is increasing abstraction and so decompression of concepts, mathematics teaching demands the opposite process as mathematical ideas are communicated to learners. Compressed forms need to be unpacked, and in Ball et al.'s terms, this is mathematical work, and a key element of the specialised mathematics teachers need to know and be able to use. Compelling as it is, the notion of unpacked mathematics, or unpacking as a way of processing knowledge, was relatively undefined, and so open to interpretation both in research and practice. In Adler and Davis (2006) we were interested in assessment in teacher education as a window into privileged knowledge for teaching, and thus whether 'unpacking' was assessed and how. We defined 'unpacking', as a particular kind of reasoning (p. 284) which we then operationalized so as to be able to unambiguously read our empirical texts. Parker (2009) developed this framing further, with additional abstractions that enabled a reading of assessment tasks in pre-service mathematics teacher education.

A Proliferation of Languages

In describing the extensive knowledge-in-use research on MKT and the smaller body of research on knowledge produced research on MKT, I have attempted to give substance to the claim that MKT, like the knowledge and research of which it is part (mathematics education) has features of a horizontal structure, and despite attempts within strands (e.g. the QUANTUM work on 'unpacking', and the measurement research), overall the grammar is weak across the range of conceptual frames that have emerged. This substantiation however, requires further systematic study. While Lerman (2006) has discussed the plurality of theories in mathematics, and whether and how this matters, an analysis of the large number of research papers produced in the past 10 years focused on MKT and using the methodological tools developed from sociology by Lerman et al. (op cit) offers possibilities for further insight into the production of this subdomain, and with this, explanatory resources of its shape and content.

Why Does MKT Matter?

A number of studies in mathematics teacher education in Southern Africa have argued for the centrality of teachers' subject matter knowledge – that professional development focused on pedagogic content knowledge is constrained by the horizon of teachers' content knowledge (Graven 2002) and that learning mathematics for teaching through research (as advocated through the action research or teachers as researchers movement) needs to place mathematics at its centre (Huillet et al. 2011). Earlier, I noted that while most of the researchers named above would agree that mathematics teachers need to know more than 'just the content', and that there is a specificity to the mathematics they need to know and be able to use, the social fact of their diverse conceptualisations of this knowledge suggests that there would not be simple agreement or homogeneity in how these might be interpreted into curricula for teacher education. Indeed, there is contestation within the mathematics education research community, as well as between them and those in the mathematics community with interests in education, as to the strength of the boundary between mathematics per se, and its use in teaching. This is not surprising, as the development of new fields and what counts as legitimate in these, is a much a political struggle as it is epistemic (Bernstein 2000, p. 162).

And this leads to the second line of work stimulated by Steve Lerman. If what counts as legitimate MKT is both epistemic and political, then who is involved in its production begins to matter.

Internal Knowledge Production, Its Enablements and Constraints

In his mapping of the effects of policy on mathematics teacher education, Lerman (2014) shows the complex position of mathematics education as a research domain in relation to the terrain of educational policy, particularly teacher education policy in the United Kingdom. He describes the mathematics education research community as largely "identical to the mathematics educators' community" (p. 13). In Bernstein's terms, mathematics teacher educators are agents in the field of production in mathematics teacher education. They are the dominant authors/researchers of research articles related to MKT. At the same time, mathematics teacher educators are agents in the field of recontextualisation. They are the same people interpreting this work into curricula for teacher education. I take some liberty here to reflect on what this dual, internal or insider position – the mathematics teacher educator-researcher – can mean.

Lerman (2014) points to the constraints of this internal functioning in our field. If, as Lerman et al. (op cit) show, mathematics education research speaks largely the mathematics education community, then its impact or influence on policy is likely to be constrained. A similar point was made in the survey of mathematics teacher education research (Adler et al. 2005) where 'insider' research dominates mathematics teacher education research. As has been argued elsewhere, within the context of higher education, despite increasing official control of teacher education curricula, there are, nevertheless, spaces for agentic action (Parker and Adler 2005). As agents in the recontextualising field, mathematics teacher educators are in a position to influence curricula in teacher education and so open opportunities for current and future teachers to learn MKT. Interesting examples of such developments in the UK are the Mathematics Enhancement Courses for graduates who wish to retrain as mathematics teachers (Adler et al. 2014), and the Teaching Advanced Mathematics course (see www.mei.org.uk) in which Steve himself has had central role.

At a more political level, however, and as noted above, MKT is part of a horizontal knowledge structure: it offers new languages and opens new positions.

Here the positions opened are those of specialised mathematics teacher educators. We (as I too am positioned here) are creating knowledge and related positions that serve our direct self-interest. The politics of this with respect to mathematicians and their role in producing MKT has formed part of the terrain, with a number of mathematicians collaborating with mathematics educators in the production of this knowledge. Hyman Bass and his collaboration with Deborah Ball and colleagues at the University of Michigan is a good example here (e.g. Ball and Bass 2000a, b). In addition others have contested mathematics education research and researchers. The 'math wars' that unfolded in the United States of America over reform of the mathematics curriculum is most illustrative of such contests.

The politics with respect to those in the official fields is less apparent. Lerman et al. (op cit) have shown that the field of mathematics education in general does not simultaneously engage, through critical research, with the official discourses in education.

In addition to positioning with respect to mathematicians and those in the official field, there are also consequences for our pedagogy. As Bernstein argues, a Horizontal Knowledge Structure consists of an array of languages; any transmission thus entails some selection or privileging:

> The social basis of the principle of this recontextualising indicates whose 'social' is speaking ... Whose perspective is it? How is it generated and legitimated? I say that this principle is social to indicate that the choice here is not rational in the sense that it is based on 'truth' of one of the specialised languages. ... Thus a perspective becomes the principle of the recontextualisation which constructs the horizontal knowledge structure to be acquired ... [and] behind the perspective is a position in a relevant intellectual field/arena. (Bernstein 1999, p. 164)

Coming to know thus means acquiring a 'gaze', and for Bernstein, particularly where grammar is weak, this is likely to be a tacit process. As argued earlier in the paper, because it is within educational discourse, and also in relative infancy, mathematical knowledge for teaching, as a new domain, has a weak grammar. What it includes and excludes, what counts as legitimate, is a function of a particular 'social' speaking, and so a perspective, that will not necessarily be explicit to learners (in this case future or practicing teachers). Rather they will be inserted in a practice which develops a particular 'gaze' on mathematics per se, and its recontextualisation in teaching.

Conclusion

My intention in this chapter has been to work with Steve's work, and hopefully invite extensions to his influence. I have focused in on recent work that has put Bernstein's sociological tools to work to interrogate the development of mathematics education research. With this social orientation to knowledge and its production in mind, I reflected on the recent but growing domain of inquiry related to mathematical knowledge for teaching (MKT).

(continued)

Drawing from research on MKT as situated knowledge, that is, mathematics in use in teaching; and MKT as knowledge produced in teacher education practice, I highlighted MKTs weak grammar through the concept of *unpacking* or *unpacked knowledge*. I also illustrated the relatively large range of conceptual frameworks circulating in the field, despite most having their roots in Shulman's seminal work on the 'missing paradigm'.

I then turned to selections from research in mathematics teacher education in Southern Africa to argue for the centrality of subject matter as key in teacher education, both preparation and professional development. This means that, if there is a specificity to teachers' mathematical knowledge for teaching, such knowledge needs to be included in teacher education programmes. With teacher educators as both the producers of such knowledge and then its recontextualisation into practice, is a danger of continuing ideological motivations driving such programmes on the one hand, and the possible dominance of implicit practices on the other. At the same time, as agents in the recontextualising field, there are possibilities for influencing and shaping teacher education productively. And this internal constraint and enablement is similarly positioned in context of increasing official control over teacher education in some, though not all countries.

A number of challenges are thus presented for our work, and my hope from this chapter, is that further work, drawing on the conceptual tools that have emerged from Steve Lerman's work, will enable us to reflexively travel this road.

Acknowledgment This paper forms part of the work of the Wits Maths Connect Project at the University of the Witwatersrand, supported by the FirstRand Foundation Mathematics Education Chairs Initiative of the First Rand Foundation the Department of Science and Technology and the National Research Foundation (NRF). Any opinion, findings and conclusions or recommendations expressed in this material are those of the author(s) and do not necessarily reflect the views of the FRF, DST, or NRF.

References

Adler, J., & Davis, Z. (2006). Opening another black box: Researching mathematics for teaching in mathematics teacher education. *Journal for Research in Mathematics Education, 37*(4), 270–296.

Adler, J., & Davis, Z. (2011). Modelling teaching in mathematics teacher education and the constitution of mathematics for teaching. In T. Rowland & K. Ruthven (Eds.), *Mathematical knowledge in teaching* (pp. 139–160). New York: Springer.

Adler, J., & Patahuddin, S. M. (2012). Recontexualising items that measure mathematical knowledge for teaching into scenario based interviews: An investigation. *Journal of Education, 56*, 1–12.

Adler, J., Ball, D., Krainer, K., Lin, F.-L., & Novotna, J. (2005). Reflections on an emerging field: Researching mathematics teacher education. *Educational Studies in Mathematics, 60*(3), 359–381. doi:10.1007/s10649-005-5072-6.

Adler, J., Hossain, S., Stevenson, M., & Clarke, J. (2014). Mathematics for teaching and deep subject knowledge: Voices of Mathematics Enhancement Course students in England. *Journal of Mathematics Teacher Education, 17*, 129–148.

Ball, D., & Bass, H. (2000a). Interweaving content and pedagogy in teaching and learning to teach: Knowing and using mathematics. In J. Boaler (Ed.), *Multiple perspectives on mathematics teaching and learning* (pp. 83–104). Westport: Ablex Publishing.

Ball, D., & Bass, H. (2000b). Making believe: The collective construction of public mathematical knowledge in the elementary classroom. In D. Phillip (Ed.), *Yearbook of the National Society for the Study of Education, constructivism in education* (pp. 193–224). Chicago: University of Chicago Press.

Ball, D., Bass, H., & Hill, H. (2004). Knowing and using mathematical knowledge in teaching: Learning what matters. In A. Buffler & R. C. Laugksch (Eds.), *Proceeding of the 12th annual conference of the Southern African Association for Research in Mathematics, Science and Technology Education* (pp. 51–65). Durban: SAARSMTE.

Ball, D., Thames, M., & Phelps, G. (2008). Content knowledge for teaching: What makes it special? *Journal of Teacher Education, 59*, 389–407. doi:10.1177/0022487108324554.

Bernstein, B. (1999). Vertical and horizontal discourse: An essay. *British Journal of Sociology of Education, 20*(2), 157–173.

Bernstein, B. (2000). *Pedagogy, symbolic control and identity: Theory, research, critique* (2nd ed.). Oxford: Rowman & Littlefield.

Davis, B. (2011). Mathematics teachers' subtle, complex disciplinary knowledge. *Science, 332* (6037), 1506–1507. doi:10.1126/science.1193541.

Graven, M. (2002). *Mathematics teacher learning, communities of practice, and the centrality of confidence*. PhD, University of the Witwatersrand, Johannesburg.

Hill, H. (2008). Unpacking pedagogical content knowledge: Conceptualizing and measuring teachers' topic-specific knowledge of students. *Journal for Research in Mathematics Education, 39*(4), 372–400.

Hill, H., Rowan, B., & Ball, D. (2005). Effects of teachers' mathematical knowledge for teaching on student achievement. *American Educational Research Journal, 42*(2), 371–406.

Hill, H., Ball, D., Sleep, L., & Lewis, J. (2007a). Assessing teachers' mathematical knowledge: What knowledge matters and what evidence counts? In F. Lester (Ed.), *Handbook for research on mathematics education* (2nd ed., pp. 111–155). Charlotte: Information Age Publishing.

Hill, H., Dean, C., & Goffney, I. (2007b). Assessing elemental and structural validity: Data from teachers, non-teachers, and mathematicians. *Measurement: Interdisciplinary Research & Perspective, 5*(2–3), 81–92. doi:10.1080/15366360701486999.

Hill, H., Blunk, M., Charalambos, Y., Lewis, J., Phelps, G., Sleep, L., & Ball, D. (2008). Mathematical knowledge for teaching and the mathematical quality of instruction: An exploratory study. *Cognition and Instruction, 26*, 430–511.

Huillet, D., Adler, J., & Berger, M. (2011). Teachers as researchers: Placing mathematics at the core. *Education as Change, 15*(1), 17–32. doi:http://dx.doi.org/10.1080/16823206.2011.580769

Krauss, S., Brunner, M., Kunter, M., Baumert, J., Blum, W., Neubrand, M., & Jordan, A. (2008). Pedagogical content knowledge and content knowledge of secondary mathematics teachers. *Journal of Educational Psychology, 100*(3), 716–725.

Lemke, J. (1993). *Talking science: Language learning and values*. Norwood: Ablex.

Lerman, S. (2006). Theories of mathematics education: Is plurality a problem? *Zentralblatt für Didaktik der Mathematik, 38*(1), 8–13. doi:10.1007/bf02655902.

Lerman, S. (2014). Mapping the effects of policy on mathematics teacher education. *Educational Studies in Mathematics, 87*(2), 187–201.

Lerman, S., Xu, G., & Tsatsaroni, A. (2002). Developing theories of mathematics education research: The ESM story. *Educational Studies in Mathematics, 51*(1–2), 23–40. doi:10.1023/a:1022412318413.

Ma, L. (1999). *Knowing and teaching elementary mathematics: Teachers' understanding of fundamental mathematics in China and the United States*. Mahwah: Lawrence Erlbaum.

Maton, K., & Muller, J. (2007). A sociology for the transmission of knowledges. In F. Christie & J. Martin (Eds.), *Language, knowledge and pedagogy: Functional linguistic and sociological perspectives* (pp. 14–33). London: Continuum.

Parker, D. (2009). *The specialisation of pecgogic identities in initial mathematics teacher education in post-apartheid South Africa*. PhD, University of the Witwatersrand, Johannesburg.

Parker, D., & Adler, J. (2005). Constraint or catalyst: The regulation of teacher education in South Africa. *Journal of Education, 36*, 59–78.

Parker, D., & Adler, J. (2014). Sociological tools in the study of knowledge and practice in mathematics teacher education. *Educational Studies in Mathematics, 87*(2), 203–219.

Rowland, T., Huckstep, P., & Thwaites, A. (2005). Elementary teachers' mathematics subject knowledge: The knowledge quartet and the case of Naomi. *Journal of Mathematics Teacher Education, 8*(3), 255–281. doi:10.1007/s10857-005-0853-5.

Shulman, L. (1986). Those who understand knowledge growth in teaching. *Educational Researcher, 15*(2), 4–14.

Shulman, L. (1987). Knowledge and teaching: Foundations of a new reform. *Harvard Educational Review, 57*(1), 1–22.

Chapter 11
Knowledge Construction: Individual or Social?

Judith A. Mousley

I well remember being at a conference of the International Group for the Psychology of Mathematics Education (PME) in Stellenbosch in 1998, listening to a plenary address given by Steve Lerman. A penny dropped. Steve had just said something like:

> The zpd is often described as a kind of force field which the child carries around, whose dimensions must be determined by the teacher so that activities offered are within the child's range. According to Davydov and followers, on the contrary the zpd is created in the learning activity, which is a product of the task, the texts, the previous networks of experiences of the participants, the power relationships in the classroom, etc. They speak of the ideas offered by one student potentially pulling other students into their zpd (Lerman 1994a). The zpd is the classroom's, not the child's. In another sense the zpd is the researcher's: it is the tool for analysis of the learning interactions in the classroom (and elsewhere). (Lerman 1998a, p. 71)

For me, this pointed to a way forward, solving a problem I had with the radical constructivist notion of knowledge being an individual commodity with the consequent implication that teachers needed to cater for the different needs of up to 30 individuals. But let me take a step back, because in Stellenbosch Steve was leading his audience and me along part of a path he had trodden previously—the part he later called the "*social turn*" (Lerman 2000a, p. 19).

First, though, let me go back in history to much earlier parts of the epistemological theory route.

J.A. Mousley (✉)
Deakin University, Melbourne, VIC, Australia
e-mail: judith.mousley@deakin.edu.au

P. Gates, R. Jorgensen (Zevenbergen) (eds.), *Shifts in the Field of Mathematics Education*, DOI 10.1007/978-981-287-179-4_11

An Historical Background

The late seventeenth and eighteenth centuries followed centuries of "dark ages" or "middle ages" repression of philosophy that was not aligned with church orthodoxy, when the Church had controlled what could be stated about knowledge—all of which was assumed to be God-given. Thus what is commonly known as "the period of enlightenment" saw both ink and blood spilled in battles between "realists" and "conceptualists", the former group seeing knowledge as perception of a given reality (after Plato), and the latter claiming it to be constructed through rational, conceptual activity (after Aristotle). Early in this period, people who argued for the latter, constructivist, position were considered to be dissidents. However, as the power of the aristocracy declined and religious revolution swept away the old order, these became times of deep-reaching unrest with potential for change. It was not long before new ideas could be received and debated more openly than they had been for many centuries.

In this period of philosophical, social, and industrial reformation, the work of John Locke (1632–1704) on "*The Understanding*" was a significant attempt to describe forms of knowledge as well as the processes by which they might materialise: *intuitive* knowledge, involving perception of fit or discord between ideas; *rational* knowledge, where other ideas are brought to the need to create fit or acknowledge discord; and *sensory* knowledge, being what is known about the existence of objects. Locke claimed that the mind might only know what it has created and proposed that meaning can only be held personally in that "*no one hath the power to make others have the same ideas in their minds*" (Locke 1689, 1 [iii] 11). He claimed that we must assume "*common consent*" about the meanings of words, but that individuals draw on a self-created "*public meaning*", appropriating certain sounds to certain ideas by "*tacit agreement*" (1689, 2 [ii] 13).

Locke's ideas were new to this historic era, and were contested (by, for example, Henry More and George Berkeley). His ideas were supported and furthered, though, by Giovanni Vico (1688–1744), who wrote about our capacity for "imaginative insight and reconstruction" (Berlin 1976, p. 107). Vico (1968) portrayed understandings as being structured from individual cognitive activity, with concepts built according to general patterns shaped by social activity (and especially linguistic interactions) in sociocultural contexts. Vico's theorising was essentially an expression of the relationships between individuals and their social contexts, and he made much of the ideas of communal wisdoms being embedded in the language of groups of people, and of discursive activity leading to the interactive, social construction of knowledge. Further, David Hume (1711–1776) published *Inquiry Concerning Human Understanding*, a set of essays abbreviating Locke's treatise but expounding further about knowledge being a product of social activity, with the potential for cognitive activity being shaped and limited by social, experiential contexts.

This emphasis on the social was reversed somewhat later in the eighteenth century, when the influential Kant (c. 1797) proposed that all knowledge is built *by individuals* out of experience through their struggle for personal rationality. This contrasted with beliefs that knowledge is deconstructed, i.e., reduced from experience, and is essentially social in nature.

These two competing positions were carried forward into the philosophical debates of the following two centuries, with developments in modern constructivism following Kant's lead and in sociocultural theory building on Hume's more social inclination.

Knowledge as Individual

In the twentieth century, first Piaget and then von Glasersfeld drew on the work of Vico, focusing on the development of knowledge as a product of individual cognitive activity—and their combined influence pushed the notion of social mediation into the background quite radically. First, schema theory, outlined by Head then by Piaget and his many followers, portrayed knowledge as personal, selective, and constructive, developed and organised as a result of interaction between individuals' sensory experience and their cognitive development (Paivio 1986). Sensory input was filtered, arranged, and stored in complex networks of concepts, rules, and strategies called *schemata* (Head 1920; Piaget 1926). These acted to shape future patterns of cognition as new experiences were filtered by cognitive activity: either assimilated into existing conceptual schemata or accommodated by a change in their structure. Both of these forms of adjustment were proposed to be aimed at equilibration of cognitive structures. Thus development of knowledge was the result of an child's own activity, facilitated by unfolding structures internal to the individual child.

Piaget's research agenda relating to mathematics learning was focused on logico-mathematical properties of individual children's behaviour and the progressive transformation of these into operational structures. Piaget also concentrated on individuals' capacity to reason—why something works or does not, why an object that is no longer visible is likely to be found in a particular area, and so on—and he believed that this reasoning capacity was age/stage related. Later, Piaget and his colleagues attended to some social catalysts for cognitive development, but the social was just that—an external catalyst for the essentially internal construction of knowledge.

Under the significant and relatively long-term influence of Piaget, the field of educational psychology became a science of the individual, and constructivism became an epistemology that saw knowledge being created by individuals, albeit in social environments. This was the scientific context for the establishment for the *International Group for the Psychology of Mathematics Education* (PME), whose name and scientific work reflected then current beliefs about mathematical learning. In fact, it was commented later that Piaget "*might be named the honorary god-father of PME*" (Ernest 1988/1990, p. 1).

The strength of this context was furthered by von Glasersfeld (e.g., 1983, 1984, 1990, 1995): often called the founder of modern "radical" constructivism. Von Glasersfeld defined radical constructivism, introducing it first to the North American group of PME, and then to the wider community with two very readable and well-argued books. He attributed the notion of constructivism to Vico, but Vico's highlighting of social experience and culture-bound knowledge were emphases that were not very evident in von Glasersfeld's writing. In particular, von Glasersfeld caused debate in the mathematics education professional community with the "basic principles" that he used to drive discussions about constructivism. He asserted that we are all "trivial constructivists" in that we all accept that knowledge "*is not passively received either through the senses or by way of communication; (but) is actively built up by the cognising subject*" (von Glasersfeld 1990, p. 23). However, he proposed that a "radical constructivist" accepts a further principle, with two parts:

> (a) The function of cognition is adaptive, in the biological sense of the term, tending towards fit or viability; and
> (b) Cognition serves the subject's organisation of the experiential world, not the discovery of an objective ontological reality. (von Glasersfeld 1990, pp. 22–23)

Von Glasersfeld articulated the difference between trivial and radical constructivism further in 1991.

> To embark on a <u>radical</u> constructivist path, thus, means to relinquish the age-old untestable requirement that knowledge must <u>match</u> the world as it might 'exist' independently of our experience ... From my perspective, those who merely speak of the construction of knowledge, but do not explicitly give up the notion that our conceptual constructions can or should in some way represent an independent 'objective' reality ... Their constructivism is trivial. (von Glasersfeld 1991, pp. 16–17)

It was this latter "radical" proposition—with an emphasis on individual, internal knowledge construction—that had become very influential in PME. In fact, there was what Kilpatrick (1987) referred to as a "*siege mentality ... a band of true believers whose credo demands absolute faith and unquestioning commitment, whose tolerance for debate is minimal, and who view compromise as a sin*" (p. 4). It is noteworthy that in 1990, the collection of papers (Davis et al. 1990) used in a national (USA) conference whose purpose was to discuss "*important issues regarding research on teaching and learning mathematics*" (p. ix) was called *Constructivist views on the teaching and learning of mathematics* and that its prime focus was on radical constructivism.

As an aside, my professional friendship group at that time called constructivism "obstructivism" because we felt that we had to adapt paper titles and abstracts to reflect the focus on individual cognition in order to have papers accepted for presentation at the PME conference.

Soon, broader discomfort with the complete backgrounding of social learning environments led to an attempt to create a compromise position, "social constructivism"; with its mathematics education focus (i.e., different from its science education focus) on how the notion of individual cognition could remain viable in the context of social group interaction. In fact, Paul Ernest introduced social constructivism to the mathematics education community as "*radical constructivism rehabilitated*" (Ernest 1988/1990) and acknowledged its intellectual origins in the writers of the time. Without refuting the idea that knowledge exists inside individual heads, Ernest described the social nature of teaching and learning contexts as well as a "new wave" of social theorists, but:

> For social constructivists, the negotiation of meanings in social situations, perhaps taking place implicitly, on a meta-level, is as important as the individual constructions of the learner. In my view, the process of that learning is not clearly elaborated. (Lerman 1998b, p. 337)

As Paul Ernest explained:

> The social constructivist thesis is that mathematics is a social construction, a cultural product, fallible like any other branch of knowledge. This view entails two claims. First of all, the origins of mathematics are social or cultural. ... Secondly, the justification of mathematical knowledge rests on its quasi-empirical basis. This is the controversial view put forward by a growing number of philosophers representing the new wave in the philosophy of mathematics, such as Lakatos (1976, 1978), Davis and Hersh (1980), Kitcher (1983), Tymoczko (1986) and Wittgenstein (1956). First, there is the powerful influence of the social context. This results from the expectations of others including students, parents, peers (fellow teachers) and superiors. It also results from the institutionalised curriculum: the adopted text or curricular scheme, the system of assessment, and the overall national system of schooling. These sources lead the teacher to internalise a powerful set of constraints affecting the enactment of the models of teaching and learning mathematics. The socialisation effect of the context is so powerful that despite having differing beliefs about mathematics and its teaching, teachers in the same school are often observed to adopt similar classroom practices (http://people.exeter.ac.uk/PErnest/soccon.htm).

Because the influence of the external, social world was recognised here, this new form of constructivism was accepted more widely. Ernest was accused of "sitting on the fence" but by 1993, he had clarified and more fully argued his position in two substantial, influential books:

> Social constructivism views mathematics as a social construction. ... Objective knowledge is internalized and reconstructed by individuals, during the learning of mathematics, to become the individuals' subjective knowledge. Using this knowledge, individuals create and publish new mathematical knowledge, thereby completing the cycle. Thus subjective and objective knowledge of mathematics each contributes to the creation and re-creation of the other. (Ernest 1991, pp. 42–43)

Thus social constructivism offered a philosophy of mathematics that allowed for the social objectivity and cultural utility of mathematics but also one that could acknowledge its fallibility and cultural limitations. Many PME members responded by focusing on the external social word's influence on the construction of knowledge in mathematics classrooms.

Constructivism remained, however, an "*epistemology that makes all knowing active and all knowledge subjective*" (Kilpatrick 1987, p. 10), and knowledge remained contained within individual minds. For example, mathematical understandings were portrayed as "mutual", "taken-to-be-shared" or "taken-as-shared" ideas; as "socially shared" or "socially distributed" cognition; and as "collective" memory (see, for example, Cobb 1990; Greeno 1991, 1997; Klein 1998; Resnick

1993; Wood and Yackel 1990; Yackel and Cobb 1996). Explanations of social development required the employment of psychological tools such as "mutual intentionality", "intersubjectivity", and "relatedness" (Newman and Holzman 1996, p. 142). In short, the idea of knowledge being individual and subjective, even if shared, was prevalent, and in 1993, Greeno noted that the psychology of individual cognition had threatened to become "*pervasive and profound*" as it had provided "*basic assumptions that underlie the organisation of our school curriculum, our assessment of student achievement, and important aspects of teachers' classroom practices*" (Greeno 1993, p. 154). However, Greeno had also observed that the notion that social experience shapes interpretation was re-emerging under the influence of "social constructivism".

That term was used, though, quite variously by different authors of Greeno's era (e.g., Bauersfeld 1992; Cobb et al. 1991; Confrey 1995; Ernest 1991, 1998, 1999; McRobbie and Tobin 1997; Solomon 1993; Steffe and Gale 1995; Voight 1994). When the term referred only to the importance of social interaction in the individual learning process, then social constructivism was not in conflict with radical constructivism and constructed understandings could still be regarded as essentially personal but socially shared. The social element to knowledge merely involved the influence of social experience in knowledge making, resulting in relatively common understandings.

> Paradoxically, the things which so clearly illustrate how different our worlds may be—language and culture—are also the things which deliver us from the danger of solipsism and which, further, enable us to construct worlds that have a substantial amount in common. . . . In any culture which manifests strong cohesion, we can even speak of the many common elements of the individual worlds as constituting the "shared consciousness" of that culture. It is this common or shared world which gives identity to the culture. (Geering 1994, p. 46)

When I joined PME in 1988, Ernest's work was very influential, "constructivism" was the buzz word and by that everyone meant either radical constructivism or at least shared consciousness, and many research reports in the scientific program were used to illustrate what "shared cognition" or its ilk looked like.

The theory was translated to practice in mathematics education research. For instance, a typical list of classroom features based on "constructivist" principles—compiled by Cobb et al. (1988)—demonstrates the individual nature of knowledge building that was so influential at the time as well as clear links with the theories of Piaget:

- Students are active participants in their own learning; constructing knowledge by reorganising their current ways of knowing.
- Children learn on the basis of their experiences. New learning occurs when current ideas are challenged by experiences and new concepts are constructed to restore coherence and meaning to experience.
- As children reflect on their activity, they reorganise their ideas at increasingly complex levels of abstraction.
- Reorganisation of ideas occurs during times of conflict or surprise or during periods of reflection and inactivity.
- Reorganisation takes place over an extended period of time and results in a feeling of inner satisfaction. (Cobb et al. 1988, p. 137)

At that stage, the concept that knowledge could be created and renewed *by* social groups or exist both *before* and *outside* of human minds in social objects and situations—a view of knowledge making that could *not* sit well as a form of constructivism—had yet not been brought effectively to the mathematics education community. Since Vico, this theory had taken a quite divergent developmental pathway.

Knowledge as Social

Continuing on from the development of hermeneutic theory about "The Understanding", some nineteenth century epistemologists (such as Hegel, Dilthey and Gadamer) developed the thesis that understanding involves interpretation where meaning of the part (e.g., a particular text under investigation) can be understood only in the light of the whole (e.g., its socio-historical context) and hence that that any personal reality cannot be known independent of its relationship with social and individual contexts as well as time. It was recognised that even understandings of the socio-historical context are socially mediated hypotheses—a product of the context of the interpreter (Hegel 1816). G. H. Mead studied with Dilthey, bringing hermeneutic hypotheses to twentieth century psychology and sociology. Mead emphasised that learning to think and understand results from participating in social interaction. He also explored how individuals take on *understandings of a group,* though "inner conversations" (Mead 1934, p. 146).

Early in the twentieth century, under the influence of productive authors such as Brouwer (e.g., 1913) and Dewey (e.g., 1916/1966), *shared activity* was acclaimed to be a principal feature of supportive learning environments as well as of the relatively recent province of institutionalised, comprehensive schooling. Notions of intersubjectivity and mutual intentionality gained attention. With regard to mathematics, it was proposed by both Dewey and Brouwer that mathematics takes place primarily in the mind as a result of human interaction. Dewey, in particular, had been influenced in the 1920s by the work of the soviet psychologist Vygotsky, and Brouwer, Dewey, and Vygotsky all wrote of knowing as "a way of *being* in the world" (Backhurst 1991, p. 85). All three researched and described interactions where the social world was primary, and all explored the interaction between social activity, social tools (particularly language), and social contexts.

Vygotsky (1978) saw child development as "*a complex, dialectical process … intertwining of external and internal factors, and adaptive processes which overcome impediments that the child encounters*" (p. 73). He believed that understanding an object (such as a word, concept, or process) requires a learner to act on it, transforming it from a social object into a subjective representation. However, his followers pointed out that this productive mediation process is not merely personal. For example,

> [The] social dimension of consciousness is primary in time and in fact. The individual dimension is derivative and secondary. ... All higher mental functions are internalised social relationships. (Wertsch 1985, p. 58, p. 66)

Here, social action was not determined by the psychological plane but by the nature of the social activity itself, and of great importance was the linkage of these interactions to macro structures and their "discourses" (Bernstein 1990, p. xxii). Vygotsky could not have been clearer with his statement that "In our conception, the true direction of the development of thinking is not from the individual to the socialised but from the social to the individual" (Vygotsky 1986, p. 32).

In fact, publications out of earlier Soviet socio-historical theory—along with the writing of Wittgenstein and other philosophers of language—had set the scene in the latter part of the twentieth century for on-going explorations of how the social environment works actively and interactively to shape knowledge. Researchers took up and developed further the hermeneutic assumption that understanding originates "*in the forms of life to which practice, in its bodily and emotionally structured form, belongs*" (van der Merwe and Voestermans 1995, p. 42). However, much of the focus remained on the individual (including how a child developed within their own "zone of proximal development", typically in PME called the zpd).

So, in summary, here we had two potential very different epistemological pathways: one with an emphasis on personal activity of the psyche in knowledge construction and the other with a primarily social dimension.

Steve had participated in the radical constructivist era, but it was knowledge about the latter route that he brought to the field, with a critique of the "shared consciousness" interpretation of "social constructivism" that had been so common and all-encompassing in PME circles. This revolution was broader than in PME though, as Bruner and Haste (1987) already had noted a quiet revolution taking place in educational psychology as a whole:

> It is not only that we have begun to think again of the child as a social being ... but that we have come once more to appreciate that through such social life, the child acquires a framework for interpreting experience and learns how to negotiate meaning in a manner congruent with the requirements of the culture. 'Making sense' is a social process; it is activity that is always situated in cultural and historical context. (p. 1)

Acknowledging Daniels' (2001) (arguably controversial) partitioning of the body of socio-historical theory with Soviet origins into (a) activity theory, (b) socio-cultural theory, and (c) situated and distributed cognition, and appreciating that others have attended to Steve's contributions to these fields, I will continue to focus on interpretations of "radical" and "social" constructivism, and particularly on Stephen Lerman's role in the field's "social turn".

The Turn to Sociocultural Theory

In his doctoral research (Lerman 1986), Steve had identified a range of theories about the nature of mathematics and considered their implications for teaching mathematics as well as for mathematics education research. In Lerman (1990), he explained how he had gleaned 42 student teachers' views of mathematics—obtained via a questionnaire and some interviews—and tried to place them on a continuum from absolutist to fallibilist perspectives. Steve wrote that, amongst other factors, views of mathematics impact on teaching practice. "*Theories of knowledge and its acquisition also provide the framework within which one views the needs for research, the role of mathematics in society and of mathematics in education*" (Lerman 1990, p. 59).

Soon after completion of his thesis, Lerman (1989) recognised radical constructivism to be "*a useful and productive hypothesis when thinking about listening to children and their mathematical learning*" (p. 211), but expressed concern about its underlying hypothesis that "*Coming to know is an adaptive process that organizes one's experiential world; it does not discover an independent, pre-existing world outside the mind of the knower*" (Kilpatrick 1987). Steve pointed out that that would imply we could never know what anyone else means and, at a deeper level, it called into question "*the nature of knowledge in general and of mathematical knowledge in particular*" (Lerman 1989, p. 211). If knowledge could only be taken-as-shared, how could there be mathematical meaning, proof, or truth? Steve noted that such questions had not been the concern of Piaget: "*his concern was not with valid or invalid mathematical statements, but with how the individual gains that knowledge*" (p. 215). In contrast, agued Steve, "*a concept is identified by its use, it gains its meaning from the shared social interpretation which is its use, and hence language, which itself is socially negotiated, and finds its meaning only in its use*" (pp. 215–216). Hence Steve declared that the type of constructivism discussed in mathematics education was "*more complete and consistent [and] empowering*" (pp. 215–216) than the intuitive form proposed by Piaget.

> If there are no grounds for the claim that a particular theory is ultimately the right and true one, then one is constantly engaged in comparing criteria of progress, truth, refutability etc., whilst comparing theories and evidence. This enriches the process of research. . . . Loss of certainty means that different theories and conjectures are comparable, examinable, and equally valid, until one establishes some acceptable criteria of 'better'. (Lerman 1989, pp. 216–217)

Thus, in 1989 Steve saw radical constructivism as a "useful" and "relativist" epistemology (p. 217). He portrayed objectivity as resting in the communal nature of knowledge, but acknowledged that language, concepts, and theories change over time and place as well as being specific to cultures.

In 1990, Steve explained the implications that the current predominant theory had for assessment:

> Such very different directions as those described by, e.g., Cobb (1986), focusing on the child's constructions, which of necessity originate in the under-standing that children bring

> into the classroom, cannot be assessed by the usual traditional methods which, in general, examine children's grasp of things taught by the teacher rather than the children's understanding. (p. 60)

This hinted at Steve's early engagement with constructivism, but this was only one theory amongst the substantial capital that he brought to the field. In 1992, Steve wrote about a "*Vygotskian radical constructivism (if such a concatenation is acceptable) [that] offers greater explanatory potential for classroom activity*" (Lerman 1992a, p. 105). He added to this concatenation with *The function of language in radical constructivism: A Vygotskian perspective* (Lerman 1992b). I can imagine today his wry grin as Steve reads his words now: "*a Vygotskian form of radical constructivism*" (p. 45); but the impact was immediate and in 1992 Hoyles included "Vygotsky" and "social/cultural perspectives" in her list of the "*multiple of approaches used within and outside PME*" (p. 281).

However, by 1994, Steve had clearly articulated how the assumptions and implications of the "*left-hand side [that] embeds the individual in social practices*" differ from those of the "*right-hand side [that] acknowledges the centrality of social interactions*" (Lerman 1994b, p. 146).

As another aside, I remember a quite serious and slightly heated discussion in a PME International Committee meeting when we discussed whether the "P" of PME needed to be changed for the Group to be more inclusive of sociological perspectives, and postmodernism in particular. Steve commented, straight-faced, "To P or not to P? That is the question". The English-speaking IC members all laughed, much to the consternation of an IC member who thought we should take this important "P" question "much more seriously". Actually, it was a lovely example of the sociocultural nature of language, literature, and humor!

Over subsequent years, Steve became a stronger leader in the field as he argued for a "*more holistic perspective of contemporary learning theory as it impacts upon mathematics pedagogy*" (Lerman and Dengate 1995, p. 26). Here, the personal concern I had with constructivism was brought to the fore:

> The role of the teacher in constructivism's most radical form is problematic. This theory claims that learners will construct unique understandings, varying in richness, whether the teacher acts as an inconspicuous facilitator with small groups of learners or as a traditional "stand-and-deliver" transmitter in a large lecture theatre. (p. 29)

Lerman and Dengate (1995) argued that, "*subjectivity is constituted in the social milieu, not arrived at by a decontextualised individual*" (p. 31), and that cognition is situated in the classroom discourse. This was articulated further by Steve in a guest editorial for *Education Studies in Mathematics* (ESM), where he focused on "*socially-situated meaning as mediating the development of consciousness*" (Lerman 1996a, p. 4).

It became clear that Steve was leading the social turn when the Journal for Research in Mathematics Education (JRME) published a seminal paper entitled *Intersubjectivity in mathematics learning: A challenge to the radical constructivist paradigm?* He wrote:

> There is a strong sense of pre-existing cultural tools, or concepts, outside of the individual's mind, in the notion of acculturation, perhaps implying a positioning of the individual within a culture or cultures, rather than the individual's autonomous construction of [his/her] own subjectivity. Similarly 'states of intersubjectivity' suggests something more than 'the role of the social dimension in these individual processes of constructing', something that takes place between people that the individual internalizes only secondarily. (Lerman 1996b, p. 137)

Here, Steve had made an overt, direct challenge to radical constructivism. In this article, Steve made a range of confronting points, including the following.

> ... the extension of radical constructivism toward a social constructivism, in an attempt to incorporate intersubjectivity, leads to an incoherent theory of learning. (Lerman 1996b, p. 133)
>
> Using the notion to slip from theory to theory, ignoring the contradictions and disagreements, on the grounds that each offers a richer explanation than the other at different times is ... to do an injustice to theory and also to be in danger of losing the coherence of each and the insights that each ... can offer. I argue here that social constructivists, for the most part, are guilty of such moves. (p. 139)
>
> To argue for an integrated social view is to argue that sometimes the filter has very large holes and what is occurring beyond the individual can somehow enter without constraint. ... it makes no sense to strengthen the functioning of the "social" into a social constructivism. (p. 140)
>
> ... no sense can be made of [subjectivities and positionings] being private (Wittgenstein, 1974), because they are internalized from cultural experiences and carried in language and ritual and other forms of essentially human communication. (p. 141)
>
> ... on the one hand, seeing the subject as separate from object, and on the other, drawing on the notion of the object as extension of the subject; on the one hand, seeing language as individual and secondary to intelligence, and on the other, allowing that ideas are carried in language, which is first on the social plane, in context-specific ways. ... attempts to incorporate intersubjectivity into radical constructivism make it an incoherent theory. (p. 148)

Calling on Vygotsky's theory, where "*there is no separation; mediation by materials, tools, peers, and teachers are constitutive of learning*" (p. 147), Steve described three forms of intersubjectivity and demonstrated how each is essentially social. They were (a) subjectivity constituted through social practices; (b) cognition as situated in practices; and (c) mathematics as cultural knowledge. Quite aggressively—for Steve—he suggested that "*mathematics education would benefit from abandoning constructivism as a view of how people learn*" (p. 134) in order to focus more on the interface between the individual and culture and on learning as a social activity.

Now here was an argument worth watching: new grist for the mathematics education community's mill! Some thought this a long-overdue shove for PME and the wider mathematics education community because language and science education theorists and sociologists seemed to have moved on to poststructuralism while mathematics education had been bogged down with structuralism. In response, some theorists (e.g., Salomon and Perkins 1998) proposed that the suggested dichotomy was an artificial one, warning that the baby should not be thrown out with the bath water as individual learning cannot be reduced to social learning, and vice versa. These authors claimed the differences lay in matters of degree, and levels of analysis; arguing that each needs to be considered as an entity in its own right, but that we need to ask how individual and social learning relate to each other. The titles of many papers of the time used metaphors such as "binoculars" and "lenses", implying that it was really only a matter of focus of the gaze or that two eyes could see things differently.

The most comprehensive and erudite response was *Interaction or Intersubjectivity? A Reply to Lerman*. Here, Steffe and Thompson (2000) questioned Steve's interpretation of Piaget, von Glasersfeld, and Vygotsky, and also argued that they had always found substantial compatibilities between the work of Vygotsky and that of von Glasersfeld. Rather than intersubjectivity, Steffe and Thompson chose to focus on "*interaction*" *as their core hypothesis: "interaction enters into radical constructivism at its very core*" (p. 192).

> ... subject-environment interactions can engender interactions within the individual, which then can modify the interacting constructs or relationships among them. These modifications, in turn, can influence subsequent subject-environment interactions, which can engender further modifications of the individual's interacting constructs. And so on. Reflexivity between the two domains of interaction is fundamental to what radical constructivists mean by "intersubjective construction of knowledge in social interaction". (Steffe and Thompson 2000, p. 193)

Steffe and Thompson still, however, debated the difference between Steve's and radical constructivism's concepts of intersubjectivity. They pointed out that Vygotsky himself had differentiated, as had Piaget,

> ... between representations that develop primarily through the operation of the child's own thought and those that arise under the decisive and determining influence of knowledge the child acquires from those around him (Vygotsky 1987, p. 173). The former "representations" certainly do not exclude the contributions of physical interactions to physical experiential reality. (Steffe and Thompson 2000, p. 198)

Pointing out that Vygotsky's later work had "*quite consciously moved beyond an explanatory framework in which speech and social interaction were seen as the sole motive force underlying psychological development (*Minick 1987*, p. 30)*" (p. 201), Steffe and Thompson concluded with the hypothetical that Vygotsky would eventually "*have accepted an approach that focuses on individual mental structures in interacting children*" (p. 201). In fact, Steffe and Thompson saw Vygotsky's later work as being subsumed and developed further by von Glasersfeld.

Criticising Steve for conflating discussion about radical constructivism and about teaching, Steffe and Thompson pointed out that, "*Researchers should not apply general models like von Glasersfeld's or Vygotsky's directly to the practice of mathematics education*" (p. 204). Never mind that constructivists had been doing this for years at PME and similar venues! But it is not my role at this stage to join the argument, so I will move on to Steve's response. I am sure I was not the only reader to turn quickly to this (necessarily) short rejoinder in the same journal issue: *A case of interpretations of social: A response to Steffe and Thompson* (Lerman 2000b).

In a response that was equally as detailed and erudite as Steffe and Thompson's, but in his typical gentlemanly way, Steve acknowledged that "*The sense of social that [Steffe and Thompson] use is developed from Piaget and von Glasersfeld and is quite coherent and strong enough for their purposes*" (p. 210) as well as that "*The explanatory power of radical constructivism and its sense of social is amply demonstrated in Steffe and Thompson's reply to my article*" (p. 210). However, he claimed that, "*There are many instances in which I think they have not understood what I wrote and others in which they misinterpret Vygotsky*" (p. 211). Steve pointed out that he did not expect agreement: "*The discussion is about two paradigms, two views of what it is to become a conscious human being, what (perhaps where) is knowledge, and what it is to learn and teach*" (p. 211). Here, Steve called on the words of Bruner:

> The two perspectives grow from different world views that generate different pedagogic strategies, different research paradigms, perhaps even different epistemologies ... Better each go their own way. (Bruner 1996)

Steve gave many examples from the two paradigms, showing where the notion of only individual construction of knowledge seemed weak. For example, regarding his key thesis of internalization, Steve wrote:

> The process is two-way. Culture and meanings are on the external plane and must be internalized by the child; they cannot be created by the child. Clearly Piaget, and Steffe and Thompson, deny this pre-existence and the possibility of its internalization, seeing it as an "illusion". These are incompatible perspectives. (Lerman 2000b, p. 213)

In 2001, Steve noted that Vygotsky's work had become better known in the mathematics education community and that now he was "*a major figure in the development of cultural, discursive psychology*" (Lerman 2001b, p. 5). Steve acknowledged his own subjectivity here.

> I happily confess that I became fascinated and excited by Vygotsky's ideas when I first came across them some 8 years ago and immediately found a strong resonance with the way in which I perceive myself to be culturally and socially situated. (Lerman 2001b, p. 5)

Steve's chapter (Lerman 2001b) outlined the main theoretical resources that he brought from outside mathematics education. This scholarly analysis clearly differentiated histories of specific ideas and metaphors, and particularly those of Piaget and Vygotsky. Most importantly, this quintessential chapter outlined sociocultural theory as it applies to "*mathematical meaning making*" (pp. 10–12)—a section that I think should be compulsory reading in all mathematics teacher education courses, backed up with a serious discussion of its implications for learning and teaching. Of course, Steve also applied sociocultural theory consistently when it comes to teacher education:

> The classroom and seminar room are complex sites of political and social influences, sociocultural interactions, and multiple positionings involving class, gender, ethnicity, teacher-student relations, and other discursive practices in which power and knowledge are situated. I believe that individualistic accounts cannot do justice to these forces. (Lerman 2001c, p. 44)

The Zone of Proximal Development

A leading point that Steve made several times in his publications of that time was that after only 11 years of publishing, Vygotsky had died relatively young, so the task remained for researchers to continue the work of developing his intellectual theories.

> Vygotsky offered, more than anything, a method for examining how people achieve consciousness and how the external becomes the internal. We have the opportunity, and the duty, to develop and extend his work and make it relevant to our times and our situations. Neo-Vygotskian work is actually Vygotskian. (Lerman 2001b, p. 211)

And indeed this is what Steve had continued to do—which brings me back to that plenary paper in Stellenbosch in 1998.

Along with many others in the PME community, because of the earlier scholarly interactions outlined above, I had been reading more of Vygotsky (along with Davydov, Leont'ev, Luria, Wertsch, Lave and Wenger, and anyone else who wrote about "activity" or "situated" learning). As a researcher, my principal interest was now in Vygotsky's notion of the zone of proximal development.

> Vygotsky identifies learning as leading development, quite the reverse of Piaget's view, and leads to his important and revolutionary notion of the zone of proximal development (Newman and Holzman 1993). For Vygotsky, learning is a process of pulling the child into her or his tomorrow rather than exercising where she or he is today. Vygotsky described learning in the zone of proximal development (ZPD) as taking place in interaction with a teacher or a more informed peer. What the child can imitate today, she or he will be able to do with assistance tomorrow and alone thereafter. These are key features of research that draws on a Vygotskian socio-cultural perspective, and they make teaching and learning an essentially integrated social activity and focus for research in education. (Lerman 1998b, p. 338)

However, in this newly "social" environment, I was not the only person who struggled with this zpd being an attribute of "the child". But here was Steve encouraging me to think of it as a *social zone*, and this suddenly enabled a huge step forward in planning my PhD research. In my mind, the role of personal zpd *led* by social zpd made sense. Further, the role of the mathematics teacher became clear: to lead expertly the setting up of the class' zpd, which would enable the leading of it forward into newer realms of knowledge. The role of a researcher was also somewhat clarified.

The social nature of the zpd along with its individual application was not a passing mention of Steve's, but an idea he articulated more deeply over the subsequent years, as the following quotations demonstrate.

> Along with Newman and Holzman (1993) I take [the zpd] to be the explanatory framework for learning as a whole [in] all socio-cultural milieus. ... It provides the framework, in the form of a symbolic space ... for the realisation of Vygotsky's central principle of development. (Lerman 1998a, vol. 1, p. 70)
>
> Internalization, the priority of the social plane, is at the heart of Vygotsky's theory, but so too is the zpd, in which what each of the participants brings to the interaction constitutes the zone. (Lerman 1990, p. 213)

> The learning activity is certainly the task set by the teacher but is also a function of the style of classroom interaction, the texts, the ethos of the school, the possibilities arising out of the particular mix of actors that day, even that moment, and so on ... The learning activity, then, ... can set up a ZPD for the participants. (Lerman 1991, p. 57)
>
> ... the zpd would be better conceptualized not as a physical space, in the sense of the individual's equipment (either cognitive or communicative), but as a symbolic space involving individuals, their practices and the circumstances of their activity. This view takes the zpd to be an ever-emergent phenomenon triggered, where it happens, by the participants catching each other's activity ... the emergence of a zpd enables the teacher and learner or the peer learners to become mutually orientated towards socially and culturally mediated meanings. (Lerman 2001a, pp. 103–104)

This clarification of the social nature of the zpd enabled me, and others, to envision a role of the teacher that clearly was very different from that of trying to scaffold the thinking of thirty individual children (and more in many other countries) as they constructed varied conceptual meanings for every specific mathematical idea, word, and process.

Here, the teacher is the initiator of children into the practices of classroom mathematics. Although of course the students do immense cognitive "work" in gaining this initiation, they are not seen as contributing to the culture of mathematics beyond making the practices of school mathematics their own. These practices focus on "scientific" concepts with the spontaneous concepts of students arising only secondarily in classrooms. (Kieren 2000, p. 230)

This hypothesis, like Steve's other conjectures about the social contexts and primacy of learning, had opened the door for sociocultural and sociopolitical discourses and practices in mathematics classrooms to be deconstructed—and hence for mathematics education research to move further down the road of epistemology.

Conclusion

It was through such extensive, authoritative discussions (whether addresses, conference papers, journal articles, or personal discussions) that Steve brought me—along with many others from around the world—from radical and/or social constructivism to sociocultural theory (and, in essence, from structuralism to poststructuralism). We all had the chance to grow *with* Steve as he generously shared his developing, scholarly, intellectual capital. Steve not only challenged individual people to run with his ideas but also dragged the top international group for mathematics education research to a position where poststructural theories were is regular use.

In discussion about this movement with Neil Pateman at PME/PME-NA in Hawaii in 2003, I remember his responding to me, "Well, PME does not do kicking or screaming, but Steve has certainly been dragging us".

In an analysis of theories used in PME, ESM, and JRME papers from 1990 to 2001 there was an increase in references to social theorists (Lerman 2010).

(continued)

The stage had been re-set for important thinking about the natures of mathematics and its learning, and even highly-respected constructivist theorists were starting to use aspects of sociocultural theory even if only as elements of a recursive, "reflexive" relationship (Cobb 2000, p. 64). In fact, by 2001 Steve was able to write, quite accurately, that:

> The theoretical fields now drawn upon by mathematics education researchers include a range of theories that take language and social practices as constitutive of consciousness, behaviour and learning. These are: social practice theory (also called situativity, communities of practice and situated cognition); sociology; and Vygotskian theories. (Lerman 2001b, p. 97)

In 2002, he traced the change in dominant theories through the special issues of ESM (Lerman et al. 2002). Further, by 2006, he could see not only the wider effects of the social turn but also a growth in constructivism:

> . . . by the insertion of new theoretical discourses alongside existing ones. Constructivism grows, and its adherents continue to produce novel and important work; models and modeling may be new to the field, but already there are novel and important findings emerging from that orientation. (Lerman 2006a, p. 9)

Continuing to lead developments, in a review of socio-cultural research in PME, Lerman (2006b), suggested key themes for the (then) current research that might appear more in the future: (a) work around the notion of identity, including mathematical identity; (b) studies of teaching mathematics, including teacher learning as well as connections between teaching and mathematical identity; and (c) who fails mathematics, and why? Steve's predictions proved right in relation to the first two research areas, but not so much yet in relation to the third.

In summary, by arguing for a more eclectic but integrated approach to theorising about cognition, Stephen Lerman's path had led us all to a richer field for theorization of both learning and teaching mathematics. Clearly the mathematics education community had had a zone of proximal development, and all that had been needed was a more informed peer to pull us into our tomorrow. Now, it seems appropriate to leave the last word to that peer.

> I am not surprised by the multiplicity of theories in our field and the debates about their relative merits, nor do I see it as a hindrance. I am more troubled by how those theories are used. Too often theories are taken to be unproblematically applied to a research study. . . . I believe that the social turn and the proliferation of social theories have enabled us to examine and research . . . issues in ways that our previous theoretical frameworks did not allow. (Lerman 2010, p. 108)

References

Backhurst, D. (1991). *Consciousness and revolution in Soviet psychology: From the Bolshevics to Evald Illyenkov*. Cambridge, UK: University of Cambridge Press.

Bauersfeld, H. (1992). Classroom cultures from a social constructivists' perspective. *Educational Studies in Mathematics, 23*, 467–481.
Berlin, I. (1976). *Vico and Herder: Two studies in the history of ideas*. London: The Hogarth Press.
Bernstein, B. (1990). *The structuring of pedagogic discourse*. London: Routledge.
Brouwer, L. E. J. (1913). Intuitionism and formalism. *Bulletin of the American Mathematical Society, 20*, 81–96.
Bruner, J. (1996, September). *Celebrating Piaget and Vygotsky: An exercise in dialectic*. Plenary address to the conference The Growing Mind, Geneva. Retrieved March 7, 2014, from the UCSC website: people.ucsc.edu/~gwells/Files/Courses.../Bruner_Piaget-Vygotsky.pdf
Bruner, J., & Haste, H. (Eds.). (1987). *Making sense: The child's construction of the world*. London: Methuen.
Cobb, P. (1986). Contexts, goals, beliefs, and learning mathematics. *For the Learning of Mathematics, 6*(2), 2–9.
Cobb, P. (1990). Multiple perspectives. In L. Steffe & T. Wood (Eds.), *Transforming children's mathematics education: International perspectives* (pp. 200–215). Hillsdale: Lawrence Erlbaum.
Cobb, P. (2000). The importance of a situated view of learning to the design of research and instruction. In J. Boaler (Ed.), *Multiple perspectives on mathematics teaching and learning* (pp. 45–84). Westport: Ablex Publishing.
Cobb, P., Yackel, E., & Wood, T. (1988). Curriculum and teacher development: Psychological and anthropological perspectives. In E. Fennema, T. P. Carpenter, & S. J. Lamon (Eds.), *Integrating research on teaching and learning mathematics* (pp. 92–131). Madison: Wisconsin Centre for Educational Research.
Cobb, P., Wood, T., & Yackel, E. (1991). A constructivist approach to second grade mathematics. In E. von Glasersfeld (Ed.), *Radical constructivism in mathematics education* (pp. 157–163, 175–176), Dordrecht: Kluwer.
Confrey, J. (1995). How compatible are radical constructivism, sociocultural approaches and social constructivism? In L. Steffe & J. Gale (Eds.), *Constructivism in education* (pp. 185–225). Hove: Lawrence Erlbaum.
Daniels, H. (2001). *Vygotsky and pedagogy*. London: Routledge Falmer.
Davis, P. J., & Hersh, R. (1980). *The mathematical experience*. Harmondsworth: Penguin.
Davis, R. B., Maher, C. A., & Noddings, N. (Eds.). (1990). *Constructivist views on the teaching and learning of mathematics*. Reston: National Council of Teachers of Mathematics.
Dewey, J. (1916). *Democracy and education: An introduction to the philosophy of education*. New York: Macmillan. (Original work published 1966)
Ernest, P. (1988/1990). *Social constructivism as a philosophy of mathematics: Radical constructivism rehabilitated?* Retrieved March 1, 2014, from the University of Exeter website http://people.exeter.ac.uk/PErnest/soccon.htm. (Originally presented as a paper, *The Impact of Beliefs on the Teaching of Mathematics*, to the *6th International Congress of Mathematical Education*, Budapest, Hungary on August 7, 1988 then revised for self-publication in March, 1990 on the university website).
Ernest, P. (1991). *The philosophy of mathematics education: Studies in mathematics education*. London: Falmer Press.
Ernest, P. (1998). *Social constructivism as a philosophy of mathematics*. Albany: State University of New York Press.
Ernest, P. (1999). Forms of knowledge in mathematics and mathematics education: Philosophical and rhetorical perspectives. *Educational Studies in Mathematics, 38*(1–3), 67–83.
Geering, H. (1994). *Tomorrow's God: How we create our worlds*. Wellington: Bridget Williams Books.
Greeno, J. G. (1991). Number sense as situated knowing in a conceptual domain. *Journal for Research in Mathematics Education, 22*, 170–218.
Greeno, J. G. (1993). For research to reform education and cognitive science. In L. Penner, G. M. Batsche, H. M. Knoff, & D. L. Nelson (Eds.), *The challenges in mathematics and science*

education: Psychology's response (pp. 153–192). Washington, DC: American Psychological Association.
Greeno, J. G. (1997). Theories and practices of thinking and learning to think (Reforming the third R: Changing the school mathematics curriculum). *American Journal of Education, 106*(1), 85–124.
Head, H. (1920). *Studies in neurology*. New York: Oxford University Press.
Hegel, G. (1816/1929, October). Logic/Wissenschaft der Logik. In P. Snyder (1929), *Logic, Journal of English and Philosophy*, 7–17.
Hoyles, C. (1992). Illuminations and reflections—Teachers, methodologies, and mathematics. In W. Geeslin & K. Graham (Eds.), *Proceedings of the 16th PME international conference* (Vol. 3, pp. 263–286). Tsukuba: PME.
Kieren, T. E. (2000). Dichotomies or binoculars: Reflections on the papers by Steffe and Thompson and by Lerman. *Journal for Research in Mathematics Education, 31*(2), 228–233.
Kilpatrick, J. (1987). What constructivism might be in mathematics education. In *Proceedings of the eleventh international conference on the psychology of mathematics education* (Vol. 1, pp. 3–27). Montreal: PME.
Kitcher, P. (1983). *The nature of mathematical knowledge*. Oxford, UK: Oxford University Press.
Klein, M. (1998). New knowledge/new teachers/new times: How processes of subjectification undermine the implementation of investigatory approaches to teaching mathematics. In C. Kanes, M. Goos, & E. Warren (Eds.), *Teaching mathematics in new times* (pp. 295–302). Brisbane: Mathematics Education Research Group of Australasia.
Lakatos, I. (1976). *Proofs and refutations*. Cambridge, UK: Cambridge University Press.
Lakatos, I. (1978). *Philosophical papers* (Vol. 2). Cambridge, UK: Cambridge University Press.
Lerman, S. (1986). *Alternative perspectives on the nature of mathematics and the possible influence on the teaching of mathematics*. Unpublished PhD thesis, Kings College, University of London.
Lerman, S. (1989). Constructivism, mathematics and mathematics education. *Educational Studies in Mathematics, 20*(2), 211–223.
Lerman, S. (1990). Alternative perspectives of the nature of mathematics and their influence on the teaching of mathematics. *British Educational Research Journal, 16*(1), 53–61.
Lerman, S. (1991). Accounting for accounts of learning mathematics: Reading the ZDP in videos and transcripts. In D. Clarke (Ed.), *Perspectives on practice and meaning in mathematics and science classrooms* (pp. 53–74). Dordrecht: Kluwer.
Lerman, S. (1992a). The position of the individual in radical constructivism: In search of the subject. In J. A. Malone & P. C. S. Taylor (Eds.), *Constructivist interpretations of teaching and learning mathematics: Proceedings of topic group 10 at the 7th International Congress on Mathematical Education, held in Quebec, Canada, August 1992* (pp. 105–112). Perth: Curtin University of Technology.
Lerman, S. (1992b). The function of language in radical constructivism: A Vygotskian perspective. In *Proceedings of the 16th PME conference, University of New Hampshire, Durham, August 6–11, 1992* (Vol. 2, pp. 40–47). Durham: PME.
Lerman, S. (1994a). Towards a unified space of theory-and-practice in mathematics teaching: A research perspective. In L. Baszini (Ed.), *Proceedings of the fifth international conference on systematic co-operation between theory and practice in mathematics education* (pp. 133–142). Pavia: Universita degli Studi di Pavia.
Lerman, S. (1994b). Metaphors for mind and metaphors for teaching and learning mathematics. In *Proceedings of the 18th international conference for the psychology of mathematics education* (pp. 144–151). Lisbon: PME.
Lerman, S. (1996a). Guest editorial. *Educational Studies in Mathematics, 31*, 1–9.
Lerman, S. (1996b). Intersubjectivity in mathematics learning: A challenge to the radical constructivist paradigm? *Journal for Research in Mathematics Education, 27*(2), 133–150.
Lerman, S. (1998a). A moment in the zoom of a lens: Towards a discursive psychology of mathematics teaching and learning. In A. Olivier & K. Newstead (Eds.), *Proceedings of*

22nd conference of the international group for the psychology of education (Vol. 1, pp. 66–81). Stellenbosch: International Group for the Psychology of Mathematics Education.
Lerman, S. (1998b). Research on socio-cultural perspectives of mathematics teaching and learning. In A. Sierpinska & J. Kilpatrick (Eds.), *Mathematics education as a research domain: A search for identity* (pp. 333–350). Dordrecht: Kluwer.
Lerman, S. (2000a). The social turn in mathematics education research. In J. Boaler (Ed.), *Multiple perspectives on mathematics teaching and learning* (pp. 19–44). Westport: Ablex Publishing.
Lerman, S. (2000b). A case of interpretations of social: A response to Steffe and Thompson. *Journal for Research in Mathematics Education, 31*(2), 210–227.
Lerman, S. (2001a). Cultural, discursive psychology: A sociocultural approach to studying the teaching and learning of mathematics. *Educational Studies in Mathematics, 46*, 87–113.
Lerman, S. (2001b). A cultural/discursive psychology for mathematics teaching and learning. In B. Atweh, H. Forgasz, & B. Nebres (Eds.), *Sociocultural research on mathematics education* (pp. 3–18). Mahwah: Lawrence Erlbaum Associates.
Lerman, S. (2001c). A review of research perspectives on mathematics teacher education. In F.-L. Lin & T. J. Cooney (Eds.), *Making sense of mathematics teacher education* (pp. 32–52). Dordrecht: Kluwer.
Lerman, S. (2006a). Theories of mathematics education: Is plurality a problem? *ZDM, 38*(1), 8–13.
Lerman, S. (2006b). Socio-cultural research in PME. In A. Gutiérrez & P. Boero (Eds.), *Handbook of research on the psychology of mathematics education: Past, present and future* (pp. 347–366). Rotterdam: Sense.
Lerman, S. (2010). Theories of mathematics education: Is plurality a problem? In G. Kaiser & B. Sriraman (Eds.), *Advances in mathematics education* (pp. 99–110). Dordrecht: Springer.
Lerman, S., & Dengate, B. (1995). Learning theory in mathematics education: Using the wide angle lens and not just the microscope. *Mathematics Education Research Journal, 7*(1), 26–36.
Lerman, S., Xu, G., & Tsatsaroni, A. (2002). Developing theories of mathematics education research: The ESM story. *Educational Studies in Mathematics, 51*(1–2), 23–40.
Locke, J. (1689). *Essay on the human understanding. In four books*. London: Clarenden.
McRobbie, C., & Tobin, K. (1997). A social constructivist perspective on learning environments. *International Journal of Science Education, 19*(2), 193–208.
Mead, G. H. (1934). *Mind, self and society: From the standpoint of a sociologist*. Chicago: University of Chicago Press.
Minick, N. (1987). The development of Vygotsky's thought: An introduction. In R. W. Rieber & A. S. Carton (Eds.), *The collected works of L. S. Vygotsky: Vol. 1: Problems of general psychology* (pp. 17–36). New York: Plenum Press.
Newman, F., & Holzman, L. (1993). *Lev Vygotsky: Revolutionary scientist*. London: Routledge.
Newman, F., & Holzman, L. (1996). *Unscientific psychology: A cultural-performatory approach to understanding human life*. London: Praeger.
Paivio, A. (1986). *Mental representations: A dual coding approach*. Oxford, UK: Oxford University Press.
Piaget, J. (1926/1950). *The psychology of intelligence*. New York: Harcourt Brace.
Resnick, L. B. (1993). Shared cognition: Thinking as social practice. In L. B. Resnick, J. M. Levine, & S. D. Teasley (Eds.), *Perspectives on socially shared cognition* (pp. 1–22). Washington, DC: American Psychological Association.
Salomon, G., & Perkins, D. N. (1998). Individual and social aspects of learning. *Review of Research in Education, 23*, 1–24.
Solomon, J. (1993). The social construction of children's scientific knowledge. In P. J. Black & A. M. Lucas (Eds.), *Children's informal ideas in science*. London: Routledge.
Steffe, L. P., & Gale, J. (Eds.). (1995). *Constructivism in education*. Hove: Lawrence Erlbaum.
Steffe, L. P., & Thompson, P. W. (2000). Interaction or intersubjectivity? A reply to Lerman. *Journal for Research in Mathematics Education, 31*(2), 191–209.
Tymoczko, T. (Ed.). (1986). *New directions in the philosophy of mathematics*. Boston: Birkhauser.

van der Merwe, W. L., & Voestermans, J. (1995). Wittgenstein's legacy and the challenge to psychology. *Theory and Psychology, 5*(1), 27–48.

Vico, G. (1925/1968). Logic. In *The new science of Giambattista Vico* (3rd ed. Revised translation of Vico's *Scienza Nuova* by T. G. Bergin & M. H. Fisch). Ithaca: Cornell University Press. (Original work published 1825)

Voight, J. (1994). Negotiation of mathematical meaning and learning mathematics. In P. Cobb (Ed.), *Learning mathematics: Constructivist and interactionist theories of mathematical development* (pp. 171–194). Dordrecht: Kluwer.

von Glasersfeld, E. (1983). Learning as a constructive activity. In J. C. Bergeron & N. Herscovics (Eds.), *Proceedings of the 5th Annual Meeting of the North American Group of Psychology in Mathematics Education* (Vol. 1, pp. 41–101). Montreal: PME-NA.

von Glasersfeld, E. (1984). An introduction to radical constructivism. In P. Watzlawick (Ed.), *The invented reality* (pp. 17–40). New York: Norton.

von Glasersfeld, E. (1990). An exposition of constructivism: Why some like it radical. In R. B. Davis, C. A. Maher, & N. Noddings (Eds.), *Constructivist views on the teaching and learning of mathematics* (JRME, Monograph no. 4, pp. 19–29). Reston: National Council of Teachers of Mathematics.

von Glasersfeld, E. (1991). Knowing without metaphysics: Aspects of the radical constructivist position. In F. Steier (Ed.), *Research and reflexivity* (pp. 12–29). London: Sage.

von Glasersfeld, E. (1995). *Radical constructivism: A way of knowing and learning*. London: Routledge Falmer.

Vygotsky, L. S. (1978). *Mind in society: The development of higher psychological processes* (from the unpublished notebooks of L. S. Vygotsky, edited and translated by M. Cole, V. John-Steiner, S. Scriber, & E. Souberman). Cambridge, MA: Harvard University Press.

Vygotsky, L. S. (1986). *Thought and language* (Revised Edition, A. Kozulin [Ed.]). Cambridge, MA: MIT Press.

Vygotsky, L. S. (1987). The development of scientific concepts in childhood. In R. W. Rieber & A. S. Carton (Eds.), & N. Minick (Trans.), *The collected works of L. S. Vygotsky* (Vol. 1, pp. 167–241). New York: Plenum Press.

Wertsch, J. V. (1985). *Culture, communication and cognition: Vygotskian perspectives*. Cambridge: Cambridge University Press.

Wittgenstein, L. (1974). *Philosophical grammar*. Oxford: Blackwell.

Wittgenstein, L. (1978/1956). *Remarks on the foundations of mathematics* (revised edition, 1978). Cambridge: MIT Press.

Wood, T. L., & Yackel, E. (1990). The development of collaborative dialogue within small group interactions. In L. P. Steffe & T. Wood (Eds.), *Transforming children's mathematics education* (pp. 244–252). Hillsdale: Erlbaum.

Yackel, E., & Cobb, P. (1996). Sociomathematical norms, argumentation and autonomy in mathematics. *Journal for Research in Mathematics Education, 27*, 458–477.

Chapter 12
Intersubjectivity in Mathematics Teaching: Meaning-Making from Constructivist and/or Sociocultural Perspectives?

Barbara Jaworski

Introduction

Intersubjectivity concerns the meanings that human beings make in relation to each other. For example, a teacher works with a group of students on some area of mathematics – let us say on *fractions*. She wants the students to make sense of fraction equivalence – for example, that (under some circumstances) 2/4 is equal to 1/2 – and what it means for fractions to be 'equivalent'. The teacher brings knowledge, vision and purpose to her teaching: for example, *knowing* about rational numbers and fraction relationships, about constructing tasks for students, and about creating a classroom environment conducive to learning; *vision* about what she wants students to understand about fractions, the meanings that they should create; *purpose* in conducting lessons, satisfying curriculum goals, preparing students for examinations and so on. To fulfil her purposes and satisfy her vision she needs to communicate with students and to use her knowledge to make that communication in some way effective in terms of what her students will subsequently know about fractions. Students, on the other hand, have their own interests, ideas and concerns. For example, they might be keen to learn about fractions and make sense of what the teacher is offering; or, they might have little interest in fractions and find the need to pay attention to fraction tasks tedious and distracting from other things they would rather be thinking about.

In everyday life, as human beings, we interact with each other, and make assumptions about common 'understandings'. Often, we take for granted that the person we are talking to will know what we are talking about and that the words we use will make sense to that person. Where we are not sure of this, we might choose our words carefully, and perhaps find different ways of expressing what we mean. We are highly sophisticated in detecting differences in meaning or nuances of

B. Jaworski (✉)
Mathematics Education Centre, Loughborough University, Loughborough, UK
e-mail: b.jaworski@lboro.ac.uk

P. Gates, R. Jorgensen (Zevenbergen) (eds.), *Shifts in the Field of Mathematics Education*, DOI 10.1007/978-981-287-179-4_12

*mis*understanding. Sometimes, however, we can remain ignorant of the (alternative) meanings that others make of what we say. Communication is not a simple process, as many scholars have observed, with bodies of theory and research devoted to its exploration.

Constructivist and Sociocultural Perspectives on Intersubjectivity

In this chapter, I want to return to an area of theory development with which I have been involved throughout my research career and which has been addressed seminally in the research and writing of Steve Lerman. I have followed and learned from the theoretical work of Steve during this time, and therefore find his writings significant to my own work. Here I focus on the concept of *intersubjectivity,* within mathematics education, in a progression from 'constructivist' perspectives on mathematics learning to perspectives which are more 'sociocultural' in nature. I shall be making reference to a small number of papers, published around the millennium, in which intersubjectivity was a 'hot' topic, but questions about the nature of intersubjectivity continue to the present day.

In the 1980s, I was introduced to the theory of *Radical Constructivism* as expressed through the scholarship of Ernst von Glasersfeld and colleagues (particularly Paul Cobb, Les Steffe and Jere Confrey in the United States and Paul Ernest in the UK) drawing extensively on the work of Piaget. Steve also (briefly) worked with this theoretical perspective (Lerman 1989) My early research, drawing centrally on the work of these scholars, was a study of the use of investigative activity in mathematics teaching and learning, together with a number of classroom teachers at secondary level, and I characterised this teaching as 'constructivist', (Jaworski 1991). Indeed, one of the teachers made the following remark "*Oh, so I was a constructivist before I knew what one was! Does that mean I constructed constructivism?*" (Jaworski 1994, p. 132). This remark hit profoundly at what it means to take a constructivist stance towards teaching.

The term 'constructivist teaching' can be seen to mean a particular style of, or approach to teaching, or it can mean the kind of teaching that emerges when a teacher takes a constructivist (theoretical) stance. Theory cannot define practice – what we do depends on many factors and does not usually fit neatly into one area of theory; also practice is not an outcome of a theoretical perspective – we cannot start from a theory and use it to form what we do. Thus, espousing constructivism as a theoretical perspective, does not bring with it particular forms of practice (such as inquiry-based activity, as it is often interpreted – Abdulwahed et al. 2012); rather, the theory offers teachers a way of planning for and reflecting on their teaching, and researchers a way of analysing and interpreting teaching. Classroom situations can be analysed through a constructivist perspective whether they involve inquiry-based activity or teacher explanation and exercises. However, it is fair to say that if one espouses a particular theoretical perspective, this might influence the way one thinks about teaching or works with learners. Von Glasersfeld suggested a number of "noteworthy consequences" of radical constructivism for educational research and teaching. One of these is expressed as follows

> The teacher will realise that knowledge cannot be transferred to the student by linguistic communication, but that language can be used as a tool in the process of guiding the student's construction. (Von Glasersfeld 1987, p. xx)

Of course, such a statement can be interpreted in many ways, and such interpretation is at the heart of what, for me, are issues relating to *intersubjectivity*.

In 1996, in JRME, Steve wrote a paper entitled *Intersubjectivity in Mathematics Learning: A Challenge to the Radical Constructivist Paradigm*. In it he focused on the nature of *intersubjectivity* from a sociocultural perspective with a critique on constructivist perspectives. He writes:

> Consciousness is constructed in communication, in discursive practices, and through acculturation. It is in this sense that Vygotsky writes of the social plane as primary both in time and fact. This is not to see individuality and differences as non-existent, nor is it to argue that the individual mind is reducible to the social plane. It is to argue that the gaze of the psychologist must be on the social and cultural practices in which meaning and purposes function so that people act in the world. ... Social settings, such as the mathematics classroom, are determined by all the actors, both present and absent, and so the intersubjectivity is a function of the time and place and the goals of the activity and the actors. (Lerman 1996, pp. 136–7)

In a response to this paper, Les Steffe (1999), from a radical constructivist perspective, wrote as follows:

> In consideration of intersubjectivity, radical constructivists make a necessary distinction between first- and second-order models .... First-order models concern the knowledge the subject constructs to organize, comprehend and control his or her experience, that is the subject's knowledge ... . Second-order models are models an observing subject may construct of an observed subject's knowledge in order to explain his or her observations (i.e. his or her experience) of the observed subject's states and activities. (Steffe 1999, p. 5)

Before going further, and perhaps to state the obvious, I will point out that these two scholars are speaking each from a theoretical perspective through which he interprets the communicative world. They are not making ontological statements as to how that world *is*. I need to remind myself of this constantly as I read their compelling arguments for why they interpret the world as they do. A danger is that the reader is drawn into an ontological state in which she or he reads what is expressed as *the way things are*; from here one can be drawn into imputations of the constructivist classroom, or indeed the sociocultural classroom, and misnomers such as 'constructivist teaching'. I will give some examples here to exemplify my argument.

Examples to Illustrate Use of Theory in Analysis of Data from Classrooms

Example 1

In *Investigating Mathematics Teaching* (Jaworski 1994), I wrote about a lesson taught by a teacher Ben, which I called 'The Moving Squares Lesson' (p. 146). Students had explored a pattern in moving squares for squares of different sizes. For example in a 4 by 4 square of 16 unit squares, the problem would be as shown in Fig. 12.1

One group of students had found, by moving squares, that the number of moves required for squares of 2×2, 3×3 and 4×4 was respectively 5, 13, and 21. Colin had conjectured that for a 5×5 square the number of moves would be 29, since he saw a difference of 8 between successive numbers of moves. I quoted from a dialogue between two students, Colin and Jenny, in which they discuss the number of moves in a 100-square, as follows:

Col *For a hundred, to work that out you have to find out 2 plus 8 plus 8 plus 8 – till you get to a hundred. What we've got to do is find a formula, so that you can just get to a hundred straight off.*
Jen *Or without adding nine on.*
Col *Without adding 8 on, yes. So that's what we're aiming towards first of all. Then we can work it out on the other ones, like 2 by 1. (Jaworski* 1994, *p. 209)*

My interpretation of the dialogue above is that Colin was seeing his pattern extending to the 100th square by adding on 8 a suitable number of times. He seemed to correct Jenny's 'nine', thinking she had said it in error, meaning 'eight'. My interpretation is that Jenny was seeing 100 as being the 10th square (10×10) and that you could get to the number of moves by adding on to the ninth square (9×9). Now, whether or not my interpretation here is correct, it can be explained by Steffe's second order models theory: I had my direct (first-order) experience of the situation which I had observed. As part of this, I tried to make sense of what Colin and Jenny were arguing. It seemed that Colin and I had different second-order models from the situation. The moment passed and was not discussed further, but it

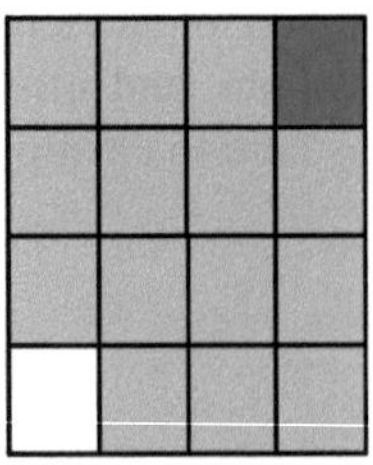

Fig. 12.1 The moving squares problem (Color figure online)

is likely that Colin and Jenny, unknowingly, went away with different generalisations of the sequence, albeit thinking they agreed with each other. Had I been the teacher (rather than an observing researcher), I might have asked each of them to explain their perceptions in order to uncover their perspectives and (perhaps) to reach some mutual understanding.

Example 2

I draw here from an example from Redmond (1992) cited in Lerman (1996, pp. 143–4).

> A researcher ... asked 5–6 year old children, in pairs, questions about "bigger" in two settings. ... [The researcher] focused on comparisons of objects, such as towers, made from building blocks, with which the children were playing. (Lerman 1996, p. 144)
>
> These two were happy to compare two objects put in front of them and tell me why they had chosen the one they had. However, when I allocated the multilinks to them (the girl had 8 the boy had 5) to make a tower ... and asked them who had the taller one, the girl answered correctly but the boy insisted that he did. Up to this point the boy had been putting the objects together and comparing them. He would not do so on this occasion and when I asked him how we could find out whose tower was the taller he became very angry. I asked him why he thought that his tower was the taller and he just replied "Because IT IS". He would go no further than this and seemed to be almost on the verge of tears. (Redmond 1992, p. 24)
>
> One can conjecture ... the inner need for the boy to assert his superiority over a girl in all things (this explanation is supported by another incident) .... (Lerman 1996, p. 144)

From a radical constructivist perspective, Steve's conjecture is his interpretation of the presented dialogue in which, from his own experience of the world of children and adults and relationships within it, he makes sense of what is reported. We might see it as a second order model in Steffe's terms. Perhaps the original researcher (Redmond) knew more about the social background of these children and was in a position to conjecture with more knowledge/experience. Steve writes, "*A radical constructivist reading might argue that in these examples individuals construe their own meanings*" (p. 144). I agree. However, we can ask on what such construal is based.

In Steve's analysis, our attention is drawn to a complexity of meanings in interpretation of the Redmond extract. Unlike in Example 1, where the analysis could focus centrally on the (mathematical) dialogue in the episode cited, Example 2 demands a wider understanding and analysis in which to embed the recorded dialogue. It is not enough to focus only on the boy's understanding of *size* in the classroom episode; the social setting in the classroom and beyond (family and society values, for example) are central to the practice or discourse within which the boy's feelings and thinking are formed. The boy's claim that his tower was bigger, was not an objective statement about the size of the tower, but a statement that was embedded in deep cultural meanings for the boy about relationships between boys and girls.

Steffe (1999) commented on Steve's analysis of this episode. He wrote,

> I note that all the interpretations Lerman made of the children's conversations were Lerman's meanings of what the children said and did ... this is the nature of second-order models. ... [T]he idea that the boy had a need to assert his superiority over a girl was an interpretation. ... But in-so-far as Vygotsky stressed that individual meanings are not simple and direct copies of social meanings, it would seem that in this case an explanation would be needed at least of how Lerman's interpretation might differ from whatever he regards as social meanings. (p. 7)

I draw attention to Steffe's imputation of a difference between *Lerman's meanings* and *social meanings*. It might be argued that Lerman's meanings *are* social meanings in so far as they coincide with meanings that are embedded in, or understood within, the communities of which Lerman is a part.

Example 3

I take my third example from research in which I engaged with Despina Potari, and have chosen the example because it captures some of the issues raised (for me) above. We worked in a small community of inquiry with two teachers who wanted to develop their practice using a theoretical construct, 'the teaching triad', consisting of three dimensions of teaching practice: *Management of Learning*, *Sensitivity to Students* and *Mathematical Challenge* (Jaworski 1994; Potari and Jaworski 2002). Our analysis of recordings of classroom dialogue was interpretative, largely restricted to interpretations within the classroom setting. Our data went beyond this setting where the teachers were concerned (through interviews) but we had no beyond-classroom data for the students we observed. We wrote two accounts from the research, one for each teacher, focusing on aspects of their practice that they considered important. In the first (Potari and Jaworski 2002), we interpreted classroom dialogue in relation to perspectives of learning and teaching mathematics, and the teacher's reflections and declared issues, using the teaching triad as an analytical tool. Although we did not declare it as such, this could be seen as a constructivist analysis (as in Example 1) in much the way Steffe (1999) describes, taking our interpretations as second-order models. In all such interpretations we took a critical stance, seeking to recognise assumptions and alternative ways of seeing events. This critique recognised elements of the social setting and the roles they might have played in the observations recorded.

The second account (Jaworski and Potari 2009) raised a range of issues which took us beyond the observations of the classroom setting in a way comparable with Example 2 above. Some interpretation was needed related to wider social issues in order to make sense of what we saw in the classroom; our data included an account of the teacher's perspective, but not that of the students. We had to conjecture the social factors in which the students' responses to the teacher were embedded. Again we might see this through the lens of Steffe's second order models. However, we offered a different story. Drawing on data from extensive observations and conversations, we characterised this teacher as an enthusiastic mathematician offering challenging tasks to his students and holding high expectations on their mathematical engagement and achievement. He recognised that in some situations students resisted his challenges, and he wished to develop his 'sensitivity to students' to become more aware of their perceptions, feelings and attitudes in order to moderate his demands on them. A range of issues emerged when we observed his teaching with a class of lower achieving students. It was hard (impossible?) to address these issues using the analytical approach we had used to date.

The teacher set his students a task with which they did not, maybe *could not* engage. It involved a homework of looking up in a dictionary, at home, the definitions of "mode", "median" and "mean" (statistical terms) and bringing their findings to the next lesson. Several students in the class had not done this task. After castigating these students for not having done their homework, the teacher handed out dictionaries so that the students could do the task there and then. It transpired that two girls thought they were looking at a French dictionary because the words they found were unfamiliar to them. The following observations come partly from our data, partly from conjecture.

- The girls had no dictionary at home.
- They were unfamiliar with finding definitions in a dictionary
- Their only experience of dictionaries was in French lessons
- They did not find the doing of homework important
- They were resentful of the way the teacher spoke to them
- They had little interest in mathematics lessons

Issues for the teacher included:

- He had prepared his lessons carefully with tasks that were designed to engage students with the mathematical topic and suitable to their levels of achievement.
- He was frustrated by some students not having attempted the dictionary task, while others had done so, since this required him, in the moment, to make alternative plans for this lesson.
- He vented his frustration by castigating students.
- He modified his task and sought to work sensitively with the students to foster their understanding of the statistical concepts.
- In seeking rapport with students, he engaged with telling facts and reducing questions to a trivial level. (Jaworski and Potari 2009)

Just as we expect teachers to have a duty of care (be sensitive to) students, researchers have similar responsibilities to all their participants including teachers. Although we might feel the teacher should have hidden his frustration through a more measured approach to his expectations of students, analysis is not about making such judgments. The teacher is a human being with human emotions. We know from our extensive discussions with him that he is in many respects a good and caring teacher; however, we also know that he realised that his challenges sometimes alienated students and recognised his own need to be more sensitive to his students' feelings and needs.

So, how did we analyse the complexity of factors and issues here? We felt the need to recognise two levels of analysis, the micro and the macro. At the micro level, we analyse the classroom observations through an interpretative process as we have described above. This has involved working on a transcript turn by turn, coding and later categorising the dialogue. This level of analysis led to a need for recognising issues which went beyond the classroom, but the classroom setting did not provide evidence to address the issues and perceived conflicts. At the macro level, we drew on school and teacher evidence to flesh out a story for the issues we identified. This resulted in our attempt to juxtapose two worlds: the world of the teacher and that of (two of) his students, and our own knowledge of society and culture in English settings. We believe, in doing so, that we went beyond the world of the classroom, with its interactions relating to mathematical topics, concepts and understandings, to the complexity of issues related to good and sensitive teaching; human emotions; low achievement; mediational tools; and home environments. In this analysis, we were not only justifying interpretations with evidence from data (where possible) but also raising questions and identifying issues, and fleshing these out in ways that we can communicate to other researchers making sense of classroom learning and teaching. In these respects we increase understandings more widely about the complexities of schooling in general, and mathematics education in particular. We used an activity theory analytical approach (based in Vygotsky, Leont'ev and Engeström) to capturing these issues and offering them to others for consideration (Jaworski and Potari 2009).

Intersubjectivity and Its Relationship with Data Analyses

I see *layers* of intersubjectivity in the examples above. The "figured worlds" of Holland et al. (1998) offer one way of making sense of these layers. A figured world is,

> . . . a socially and culturally constructed realm of interpretation in which particular characters and actors are recognized, significance is assigned to certain acts, and particular outcomes are valued over others. (p. 52)

Figured worlds can also be called figurative, narrativized or dramatized worlds, taking the form of stories or dramas by which particular events can be told, or "authored" (Holland et al. 1998, p. 53).

In each of the above examples we are introduced to some event or events, presented as a result of a researcher's analyses in a given setting. So, one of the worlds is that of the researcher. The researcher interprets data and seeks to look analytically at the other worlds that are studied. This involves interpretation and conjecture, and the rigour of the research lies in justifying interpretations and rooting conjectures. The result is some narrative that offers insights to the studied worlds. In the examples we may see worlds as follows.

In Example 1, I focus on the world of the classroom as studied by the researcher (myself) within which I focus on (albeit extremely briefly) the worlds of Jenny and Colin. The narrative concerns only their mathematical discourse in relation to the moving squares problem. I conjecture that they have differing perceptions of generality in this problem and my focus lies within the mathematical discourse. In the book (Jaworski 1994), I explore also the dialogue in their classroom group where they discuss the intentions of the teacher in setting them this kind of problem (p. 153). We see the students trying to make sense of the meanings of their teacher and disagreeing on what they perceive these meanings to be. Here, intersubjectivity is interpreted from a constructivist perspective. I compare the dialogue between Jenny and Colin and others with that in another lesson "the vectors lesson" where there had been a need to focus on conventions for expressing vectors, and dialogue had revolved around making sense of these conventions. I wrote,

> Established conventions can be seen to have arisen from social interaction and mediation. They might be regarded as 'intersubjective' knowledge – knowledge resulting from negotiation between individuals. Radical constructivism does not exclude the possibility that individual construction is influenced by social interaction, but it excludes the possibility of recognizing knowledge outside the individual. (Jaworski 1994, p. 209)

Nevertheless, later in the same page, I discuss the discourse in the vectors lesson in which students had discussed seeing a vector as 'a movement from place *a* to place *b*'. I wrote as follows:

> [Seeing a vector in this way] could be seen as part of the intersubjectivity in the classroom. The teacher and many students spoke of it and acted as if it had shared meaning. Now, from a radical-constructivist perspective, this would be seen as all individuals having their own independent perceptions of this metaphor. However, through the discourse, these individual perspectives were negotiated, and the language involved was interpreted and construed. It could therefore be seen as if a common or intersubjective understanding developed, as if there was meaning in the classroom which was a direct product of interaction. (p. 209)

This idea of common meaning is at odds with the central principles of radical constructivism (Von Glasersfeld 1990). However, the idea of common meanings has also been recognised, from a radical constructivist perspective, by other scholars who talk about "taken-as-shared meanings" (Cobb et al. 1993, p. 26) or "agreeing to agree" (Confrey 1995, pp. 217–8). The issue here is that radical constructivism denies the possibility of knowledge being recognised as external to individual cognition whereas an acknowledgement of some form of common knowledge suggests that people have the same second-order meanings. Hence the need to say "taken-as" shared: we cannot know whether indeed they are shared.

In Example 2, the worlds of two young students in a classroom, as portrayed by a researcher, Redmond, are presented and discussed by Steve. We consider here the worlds of the two students and the two researchers. Redmond presents the narrative of the two children. Steve analyses the classroom situation from Redmond's narrative. We might see the layers of interpretation resulting in (taken-as-) shared meanings. Steve talks about 'multiple subjectivities'.

> The interpretation of the "objective" questions by the researcher and children are shared by virtue of a school mathematics setting being called up by the tasks for participants . . ., but the subjective questions call up a different practice for this boy. One can conjecture what that practice might be: the inner need for the boy to assert his superiority over a girl in all things . . . . Thus it is not appropriate to talk of a child "understanding size", nor indeed lacking an understanding; such judgments are far too simplistic and rely on a single fixed meaning for an activity such as "comparison of size". "Understanding" size takes different forms in different settings. (Lerman 1996, p. 144)

From what I see in Steve's narrative of this example, I agree with his conjecture. I recognise the different meanings within a society with which I am familiar. So far we are interpreting what we are told of the event; making sense of it, and it seems as if these meanings are shared (we can 'take' them 'as' shared). However, Steve goes further to consider a theoretical perspective on the analysis.

> A radical constructivist reading might argue that in these examples individuals construe their own meaning. I want to argue that subjectivities are regulated within and through the discourses, or practices, carrying the "force" mentioned above. (p. 144)

I think this is a fair comment about a radical constructivist reading. I am not sure that the remark about subjectivities contradicts this reading. However, I want to take this up in the next example where I will also address the notions of social meaning and of "force".

In Example 3, we can recognise worlds of the researchers, the teacher, and the students in his class. It seems clear that the teacher and his students make meaning in the dictionary episode in quite different ways, deriving from their personal knowings in differing worlds. In Steve's terms we might speak of different *subjectivities*. The teacher is rooted in his worlds of mathematics, didactics and pedagogy. He designed innovative tasks for his students to enable them to learn mathematical concepts, but he does not see this from the perspectives of the students, who are rooted in their own social worlds of friends, family and aspects of society related to their own circumstances. We conjecture that, in these worlds, it is neither common to have a dictionary at home, nor to be familiar with looking up English worlds in a dictionary to discover their meaning. Thus, for them, the teacher's task is undoable and they are upset by his castigation of them which appears very unfair. In the dialogue which follows, teacher and students move from a position of conflict to an uneasy working relationship and, eventually, both try to overcome this setback by interacting in a familiar classroom manner – the teacher encourages the students to take part in classroom activity, and moderates his challenges to enable them to do it; the students gradually engage in the new task and try to do what the teacher wants of them (Jaworski and Potari 2009).

It is possible to analyse the above from a radical constructivist position. Both teacher and students make sense of the situation from within their own worlds. The way each interprets it is his or her own construction which then is the 'force' for their resulting behaviour. However, this analysis foregrounds the sense-making of the individuals and backgrounds the social forces which lead to the ways of being and doing that we see. Regarding 'subjectivities' Steve writes, *"[T]he individual is more appropriately thought of as constituted through multiple subjectivities, a fragmented self rather than an autonomous unitary subject*" (Lerman 1996, p. 143). If we see the 'force' for these subjectivities being the social and cultural worlds of which we are a part, then we see the teacher coming from a world in which homes have dictionaries and it is common to look up the meaning of a word, and the particular students coming from homes which do not have dictionaries, so that practice involving dictionaries are alien to them. These cultural origins exert a strong force on these people to act in the ways we have seen. A sociocultural position foregrounds the social, and backgrounds individual mentality, as in the familiar quotation from Vygotsky (1978, p. 57) that, "Every function of the child's cultural development appears twice: first, on the social level, and later, on the individual level; first, *between people (interpsychological),* and then *inside (intrapsychological)*". Wertsch (1991), drawing on Vygotsky and Leont'ev, writes that rather than "the idea that mental functioning in the individual derives from participation in social life", "the specific structures and processes of intramental processing can be traced to their genetic precursors on the intermental plane" (p. 27). Such foregrounding and backgrounding, with the associated force it brings to analysis, is central to a choice of theoretical position for analysis. In our analysis in Example 3, despite a constructivist perspective offering a way of theorizing what we observed, we needed to acknowledge the particular social forces which influenced the ways in which teacher and students responded to each other. Juxtaposing aspects of these alternative worlds using an activity theory perspective (drawing on Vygotsky, Leont'ev and Engeström) provided a meaningful account which emphasized issues and went some way towards explaining the conflict. Thus in this case a sociocultural lens proved more useful than a constructivist lens.

It is clear therefore, that constructivism and sociocultural theory are two alternative perspectives and finding a compromise between them, such as a 'social constructivism' does not make sense. Steve writes, "*Either there is no force to cultural knowledge and discursive practices, or there is*" (1996, p. 140). This statement seems to capture the essential difference. Ultimately, a unit of analysis is needed which allows researchers to analyse the situation observed and make meanings from them. I spoke above of the decision of Potari and myself to analyse data from both micro and macro perspectives. In doing so, it did not make sense to use a constructivist lens. However, we might ask whether there are situations in which a constructivist lens is not only adequate but appropriate.

In analysing dialogue which focuses on mathematics, researchers may wish to trace the development of mathematical concepts by students from the students' ways of expressing the mathematics they consider. This was the case in Example 1. From the interactions of Colin and Jenny, focusing on the moving squares problem, it was possible to trace their developing mathematical thinking. In doing so, we (the teacher and I) could learn something about these students' thinking. Unlike the situations in Examples 2 and 3, it seems reasonable to study the Example 1 dialogue from within the classroom culture, taking a constructivist perspective. This allowed us to focus on mathematical sense-making *per se* to produce a meaningful account of students' thinking. In this case there seemed to be no 'force' from cultural considerations.

Two Theories

In a short chapter it is impossible to address all the theoretical nuances of the debate between radical constructivist and socio-cultural theorists. I have emphasised above the importance of these theoretical perspectives to analysis of data from mathematics classrooms. The theory acts as a lens into the situations that are analysed. In fact Steve has written about the "zoom" of a lens as a metaphor to capture differing degrees of emphasis (Lerman 1998). Steffe and Thompson (2000, p. 197) write:

> [L]earning how to engender collaboration in children's mathematical interactions as well as how to sustain and modify the collaborations in such a way that one can predict at least in broad terms how they are individually affected by the interaction is essential. In radical constructivism, one cannot take intersubjectivity as given – as an explanatory concept in the construction of knowledge – as Lerman wishes to do. Rather intersubjectivity must itself be explained.

In contrast Steve writes:

> I am not denying the function of the individual's interpretative framework, as will be obvious throughout this article. There is a dialectic of thought and language, however, whereby the former is what the individual brings to the interaction and is compatible with Steffe and Thompson's theories, but the latter is not compatible because it is what pre-exists the individual and is external to her or him. . . . Of course negotiation that can be described, to some extent, in Steffe and Thompson's terms as reciprocal interaction through assimilation goes on between people. From the point of view of cultural psychology however this negotiation presupposes a common form of life that is established through social rule following. (Lerman 2000, p. 214)

I have used these two quotations to capture an aspect of the debate that has been important in analyses such as those referred to above. The idea of 'common knowledge' as something outside of individual consciousness is clearly a problem for radical constructivism (hence the language of 'taken-as' shared). Steffe's second order models go some way to addressing the differences that can be seen in individuals' perspectives in a situation, and are a mechanism for 'explaining' intersubjectivity. Steve acknowledges 'what an individual brings to the interaction', but highlights importance of 'social rule following' for which second order models cannot account. In Examples 2 and 3, the conflicts that occur can be seen as due to social rule following to which individuals concerned do not have access and hence cannot be a part of their second order models.

The theories are not interchangeable and their approaches to conceptualising knowledge construction are incommensurable. This says nothing about whether they are right or wrong. What is clear, is that analysts need to be extremely transparent in their use of theory so that their arguments are rigorously justified in relation to their data and narratives deriving from the data.

To Conclude

Although I have drawn substantially on a debate which took place more than 15 years ago, these theoretical issues are, in large part, still relevant today. I work in a Mathematics Education Centre (www.lboro.ac.uk/departments/mec/) in which we study the learning and teaching of mathematics from both cognitive and sociocultural perspectives (leading to fierce debate at times) – and Steve is one of my colleagues. So, perhaps it is unsurprising that this debate still has considerable currency.

References

Abdulwahed, M., Jaworski, B., & Crawford, A. R. (2012). Innovative approaches to teaching mathematics in higher education: A review and critique. *Nordic Studies in Mathematics Education, 17*(2), 49–68.

Cobb, P., Yackel, E., & Wood, T. (1993). Theoretical orientation. In T. Wood et al. (Eds.), *Rethinking elementary school mathematics: Insight and issues* (JRME monograph, no. 6, pp. 21–32). Reston: National Council of Teachers of Mathematics.

Confrey, J. (1995). How compatible radical constructivism, sociocultural approaches, and social constructivism? In L. P. Steffe & J. Gale (Eds.), *Constructivism in education* (pp. 185–225). Hillsdale: Lawrence Erlbaum.

Holland, D., Lachicotte, W., Skinner, D., & Cain, C. (1998). *Identity and agency in cultural worlds*. Cambridge: Harvard University Press.

Jaworski, B. (1991). *Interpretations of a constructivist philosophy in mathematics teaching*. Unpublished PhD thesis, The Open University, Milton Keynes.

Jaworski, B. (1994). *Investigating mathematics teaching: A constructivist enquiry*. London: Falmer Press.

Jaworski, B., & Potari, D. (2009). Bridging the macro-micro divide: Using an activity theory model to capture complexity in mathematics teaching and its development. *Educational Studies in Mathematics, 72*, 219–236.

Lerman, S. (1989). Constructivism, mathematic and mathematics education. *Educational Studies in Mathematics, 20*, 211–233.

Lerman, S. (1996). Intersubjectivity in mathematics learning: A challenge to the radical constructivist paradigm? Journal for Research in Mathematics Education, 27(2), 133–150.

Lerman, S. (1998). A moment in the zoom of a lens: Towards a discursive psychology of mathematics teaching and learning. In A. Olivier & K. Newstead (Eds.), *Proceedings of the 22nd conference of the International Group for the Psychology of Mathematics Education* (Vol. 1, pp. 66–81). Stellenbosch: University of Stellenbosch.

Lerman, S. (2000). A case of interpretations of *social*: A response to Steffe and Thompson. *Journal for Research in Mathematics Education, 31*(2), 210–227.

Potari, D., & Jaworski, B. (2002). Tackling complexity in mathematics teacher development: Using the teaching triad as a tool for reflection and enquiry. *Journal of Mathematics Teacher Education, 5*(4), 351–380.

Redmond, J. (1992). *Are 4–7 year-old children influenced by discursive practices when asked to make comparisons using quantities?* Unpublished manuscript, South Bank University, London.

Steffe, L. P. (1999). Intersubjectivity in mathematics learning: A challenge to the radical constructivist paradigm. A reply to Lerman. *Chreods, 13*, 1–16.

Steffe, L. P., & Thompson, P. (2000). Interaction or intersubjectivity? A reply to Lerman. *Journal for Research in Mathematics Education, 31*(2), 191–209.
Von Glasersfeld, E. (1987). Constructivism. In T. Husen & N. Postlethwaite (Eds.), *International encyclopedia of education* (Supplement, Vol. 1). Oxford: Pergamon.
Von Glasersfeld, E. (1990). An exposition of constructivism: Why some like it radical. *Journal for Research in Mathematics Education*. Monograph, Vol. 4, Constructivist views on the teaching and learning of mathematics, pp. 19–29 and 195–210.
Vygotsky, L. (1978). *Mind in society: The development of the higher psychological processes.* London: Harvard University Press.
Wertsch, J. V. (1991). *Voices of the mind.* Cambridge, MA: Harvard University Press.

Chapter 13
Learning as Participatory Transformation – A Reflection Inspired by Steve Lerman's Papers and Practice

João Filipe Matos

Constructions of reality, ideologies, world views, are not merely alternative theories with metaphysical implications only, to be discussed, compared, refuted, or supported in the ivory towers of philosophy. Knowledge and Power are inseparably linked, and knowledge is used as and for power, the domination of one group over another, the oppression of people, the legitimation of that oppression and the rationalization of values.

(Lerman 1996, 1992, p. 173)

Introduction

In his seminal paper *The Social Turn in Mathematics Education Research* (Lerman 2000a) Steve Lerman provides examples of indicators of the receptivity of the mainstream mathematics education community to social theories of mathematics learning. In that he also stresses the idea that perhaps the reception of that article was due more to political concerns that inequalities in society were reinforced and reproduced by differential success in mathematics, than to social theories of learning. He suggests (Lerman 2000b):

> the greatest challenge for research in mathematics education (and education/social sciences in general) from perspectives that can be described as being within the social turn is to develop accounts that bring together agency, individual trajectories, and the cultural, historical and social origins of the ways people think, behave, reason and understand the world. (Lerman 2000b, p. 368)

In parallel Steve acknowledges that the work of Vygotsky and Vygotskian researchers represented a growing source of theoretical inspiration in mathematics education research. Now 15 years after the appearance of that article, the influence of Steve's theoretical elaboration in a variety of topics and issues in mathematics education research is still apparent.

J.F. Matos (✉)
Institute of Education, University of Lisbon, Lisbon, Portugal
e-mail: jfmatos@ie.ulisboa.pt

P. Gates, R. Jorgensen (Zevenbergen) (eds.), *Shifts in the Field of Mathematics Education*, DOI 10.1007/978-981-287-179-4_13

In several publications Steve Lerman discusses the developments of situated learning theories and their relevance to mathematics education research. For example, in 1998 (Lerman 1998) he critically acknowledged and discussed the revolutionary notion of the situatedness of knowing, meaning and acting and the centrality of learning from Lave and Wenger (1991) and then 10 years later made a deep and challenging analysis and discussion of the concept of a community of practice (Kanes and Lerman 2008).

Steve played a major role in my own particular trajectory in mathematics education research both in inspiring and in stimulating my work, but also in encouraging me to take risks. The continuous interaction I had with Steve for 3 years in a Portuguese funded project on Mathematical Thinking, made it possible for me to understand the rigour, the extremely deep scientific honesty and the creativity shown by Steve.

In this chapter I pursue a reflection and discussion started in 2010 and strongly influenced by several papers (some on them crucial, such as Lerman 2006) and many moments of discussion with Steve. Putting the focus on mathematics learning, I discus possibilities of a conceptual articulation between theoretical views with rather different socio-historical developments and the relevance to mathematics learning research. I draw on previous work I have undertaken with Madalena Santos (Santos and Matos 1998, 2000, 2008) and from more recent developments (Matos 2010).

A considerable body of work has been and is still being produced within the international mathematics education community focusing on learning in practice, ranging from professional education and development of teachers of mathematics (e.g. Even and Ball 2009; Matos et al. 2009) to mathematics learners in transition in a variety of contexts of mathematical practices (e.g. Abreu et al. 2002; Matos et al. 2002). In several papers published in a variety of journals and book chapters during the last 15 years, Steve Lerman discusses what constitutes the rationale for the 'social turn' and points to three main intellectual resources (Lerman 2001): anthropology (as situated theories of communities of practice), sociology and cultural/discursive psychology with its roots in Vygotsky's theories. His brief review (Lerman 2001) leads to the need to encompass in a more significant form, a sociological orientation.

My intention in this chapter is to contribute to the analysis and discussion of social theories of mathematics learning, articulating and putting in dialogue a view of learning-in-practice taking in particular the notion of learning as participation (drawing on situated learning theoretical perspectives) and learning as transformation (drawing on expansive learning perspective in activity theory). I will concentrate on discussing and exploring the very idea of "*learning as an integral part of generative social practice in the lived-in world*" (Lave and Wenger 1991, p. 35) positioning subject and community within a system of activity (Engeström 1987), acting upon (conceptual) objects and transforming them into mathematically acceptable and legitimated outcomes. In doing so I hope to contribute to the discussion and analysis challenging the possibility of conceptual dialogue between

Jean Lave and Etienne Wenger's social theories of learning and Yrgo Engeström's activity theory-based ideas on expanded learning (Engeström 2007).

First, I should start by underlying, as we have in Santos and Matos (2008), that there are common roots between the situated perspective of Lave and Wenger (1991) and the activity theory as it is recognized by those authors and explicitly referred for example by Engestrom and Cole (1997) and Miettien (1999). Jean Lave finds in the socio-cultural approach (and in particular in activity theory) key issues that serve the way she addresses activity:

(i) a way of conceptualizing activity that makes possible the analysis of its intrinsic organization through the definition of a categorization of levels of activity but that simultaneously recognizes and considers its holistic nature and dynamics;
(ii) the emphasis given to the relational nature of meaning (localized in the relations between the different levels of activity) and the activity system (which operates between the levels of activity such as in the interface action-operation); and
(iii) the dialectic analytical approach to activity and its meaning in the relations constitutive of the activity system.

Secondly, the concept of social practice is more and more explicit in the work of activity theory. For example, Toulmin (1999) elaborates on the idea of knowledge and makes a comparative analysis of the epistemological ideas of Vygotsky and late Wittgenstein. Here he identifies in both a concern with the concept of practice and that 'practice' should be the key central notion in wherever new theory of knowledge is developed in the future. In the Foreword of the book *Activity Theory and Social Practice*, Hedegaard et al. (1999) give an account of how the concept of social practice is brought into the context of the discussion of activity theory. Furthermore they identify the relations that seem to be possible between 'social practice' and 'activity' indicating implications for further research in the area aiming to clarify those relations. As Santos and Matos (2008) indicate, although Hedegaard et al. (1999) underline the importance of the concept of practice as it *"provides a way to characterize those aspects of social practice that are believed to provide the conditions for psychological development"* (p. 19) they recognize the need to think more profoundly about the possibility of existing wider meanings of social practice that exceed the notion of activity. According to Jensen (1999) activity theorists

> have not applied their insight about the situated nature of practice and the practice-situatedness of concepts reflexively, only rarely have activity theorists accounted for their own concepts and theories as embedded in activities and practices. (p. 84)

Concepts in Action Involved in the Idea of Learning-in-Practice

Taking "*learning as an integral part of generative social practice in the lived-in world*" (Lave and Wenger 1991, p. 35) means that participation in social practices does not merely influence otherwise autonomous cognitive processes. From this point of view, learning means changes in the ways that a person *participates* in social practices. Therefore, understanding how learning occurs and what is involved in learning mathematics, implies addressing the social practices where learners are engaged as it is the "*primary, generative phenomenon and learning is one of its characteristics*" (Lave and Wenger 1991, p. 34). This fundamental idea takes as crucial the "*integral nature of relations between persons acting (including thinking and learning) and the social world, and between the form and content of learning-in-practice*" (Lave 1997, p. 20) identified in the work of activity theorists such as Engeström (1987, 2001, 2007) and Davydov and Radzikhovskii (1985).

Santos and Matos (1998, 2002) reported, from a situated learning point of view, how the concepts of artefact and resource were useful in the analysis of learning. The idea of artefact is quite strongly used in Activity Theory in parallel with the notion of *tool*. The concept of resource is explicitly used by Lave and Wenger (1991) in a situated approach to learning. However, the discussion of the social nature of the human construction of mathematical artefacts deserves additional attention as suggested by Lerman (1994).

Activity in Activity Theory

Within the conceptual framework of dialectic materialism, the notion of activity is an initial abstract (Davydov 1999)., Ilyenkov was one of the authors who contributed to the development of this notion, through reflection on the relations between activity and consciousness – as the reflection of subjects on reality, their own activity and themselves–addressing consciousness as co-knowing. According to Bakhurst (1997), Ilyenkov elaborated on a theory of the ideal in which activity becomes literally part of the mind as the ideal constructs the ability to think. He goes on to argue for the capacity to act in accordance with what is proper in a cultural context and therefore he identifies thinking as a kind of activity. Hence, activity is no more seen as an abstraction but as the basic unit of analysis of consciousness (Santos and Matos 2002).

Leont'ev (1978) presents an approach to the concept of activity drawing on the idea of the structure of activity with several components establishing basic categories of human activity and allowing the possibility of researching the way individual consciousness is organized through particular and specific activities – the analysis of activity abstracted from the forms it takes while running. One basic principle for Leont'ev is the recognition of the social and cooperative nature of human activity.

He assumes human individuality is emergent from social activity thus conducive to the need to reflect upon the relation between individual consciousness and the specific activities. He sees activity as a molar unit, not an additive one in the life of the person, but a system with its own structure, its transitions and its internal transformations, and importantly its own development. He identifies non-additive elements linked to central concepts: activity (linked to motives), action (linked to a goal) and operation (linked to conditions). The motives of the activity are intimately connected to the needs felt by the individual, thrown to the activity as a form of responding to those needs. Activity involves different actions aiming to specific results intimately related to the activity and, in this way, directing the activity. Action can be made concrete in different ways and forms through operations according to the conditions available but always aligned with the goal that is supposed to be achieved.

Two key implications emerge from the approach to activity proposed by Leont'ev which are coherent with the idea of learning-in-practice: (i) activity cannot be reduced to a set of simpler stand-alone additive parts or processes, and (ii) its structural and functional unit can only be examined looking at the phenomenon in its active or live state. Goal and motives are the peculiarities of a given activity that allow us to distinguish one activity from another. But those elements have only a potential character in the activity; they are neither deterministic nor definitive as activity can only be realized through development that involves transformations given its dynamic nature.

Artefacts in Activity Theory

Taking the model of the structure of the system of activity proposed by Engeström (1999) I will concentrate on one of its elements – the artefacts, in relation to other elements addressing in particular the idea of mediation that is one of the key concepts of the socio-historical-cultural approaches. The concept of *artefac*t attracts a variety of researchers in mathematics education and is frequently used, for example, in studies that focus on the use of digital technology in education. It is common to see research considering the notion of artefact in two different forms. On the one hand artefacts are referred as tools and signs that mediate action and on the other hand we find researchers who consider external (or physical) artefacts and internal (conceptual or cognitive) artefacts. The key issue is that in both approaches it is the internal character of the artefact that makes its classification, independently of the kind of activity where the use of the artefact takes place.

Engeström (1987) considers that none of those highly dichotomized forms of conceptualizing artefacts is useful and discusses that problem in the context of the non-definitive nor rigid nature of activity. In activity, functions and uses of artefacts are in constant dynamic transformation. Elements that seem to be internal in a certain moment are externalized (for example through speech) as much as the external processes in certain occasions can be internalized. Freezing and splitting those processes seems to be a poor basis to understanding artefacts and their role in activity (Engeström 1999). The functions and use of artefacts are in a constant fluidity and transformation that goes along with unfolding activity. In this sense, the artefacts are not something fixed and external to the practices but are constitutive of practices; its usefulness is not revealed in the characteristics identified independently of its use in the practices where they are put in action. Artefacts are artefacts-in-the-practice should be understood in interaction with the forms of use that users develop in those practices.

Engeström proposes a differentiation in regard to the uses of artefacts:

> The first type is **what** artefacts, used to identify and describe objects. The second type is **how** artefacts, used to guide and direct processes and procedures on, within or between objects. The third type is **why** artefacts, used to diagnose and explain the properties and behaviour of objects. Finally the fourth type is **where to** artefacts, used to envision the future state or potential development of objects, including institutions and social systems. (Engeström 1999, p. 382, emphasis in the original)

This original classification highlights that an artefact is not considered by itself in isolation instead it is conceptualized in relation to a specific use and is always inserted in a system of activity. As Engeström points out above, the construction and transformation of (conceptual) objects mediated by artefacts is a collaborative process in its nature and dialectic in its core and where different perspectives and voices meet, collide and mix. This framework gives visibility to some characteristics that draw on the collective essence of activity – and thus potentially conflicting – not isolated nor harmonious.

A Situated Perspective and the Concept of Activity

In analysing shopping at the supermarket, Lave et al. (1984) ask "*what is it about grocery shopping in supermarkets that might create the effective context for what is constructed by shoppers as 'problem solving activity'?*" (p. 68). Grocery shopping is seen as an activity that occurs in a specialized setting designed to support it (the supermarket) constituting the arena of grocery shopping as an institution at the interface between consumers and suppliers of grocery commodities. This way of conceptualizing the relations between two layers of activity: grocery shopping and problem solving – shopping helping to shape problem solving through the setting intentionally. It may be thought of as the locus of articulation between the structured arena and the structured activity. Theorizing about the interrelations between activity and setting, Lave et al. (1984) recognize the value of the conceptualization of the idea of a setting but go against a unidirectional, setting-driven relation between activity and setting. This would reduce activity to a passive response to the setting precluding the analysis of the internal relationships within the activity. It is, though, pertinent that Lave recognizes the way activity theorists conceptualize the idea of activity as system with structures, internal transformations and self-development as it allows and creates a basis to the study of the intrinsic organization of activity. It is also recognized that the studies of Zinchenko of the holistic nature of activity (developed in the framework of activity theory) help to support the idea that understanding the nature of learning in mathematics requires a contextualized understanding of its role within that activity. This is a strong argument of the need for analysing any segment of activity in relation to the flow of activity of which it is a part.

Another relevant aspect of activity theory that deserves the attention and reflection of situated learning authors is the relational emphasis of the activity theorists underlining the parallelism found in the distinction made by Leont'ev between (personal) sense and (public, societal) meaning and the distinction Lave proposes between the constructs of (personal) setting and (public, non-negotiable) arena. In addition, the dialectic character of the analysis of activity is central to the situated perspective assumed by Lave (Santos 2004).

Lave (1996) uses the term '*ongoing activity*' to refer to activity and this form of talk orientates our attention to the strongly fluid and dynamic character of activity. It induces the interrogation of the continuity and of trajectories within the activity – where does it come from, where does it go? This relates to the holistic but local character of activity with the resources, the constraints and the actors in place present in the situation. The ongoing character of activity introduced by Jean Lave seems to be consistent with the view of Leont'ev about the notion of activity that took him to defend that it should be analysed in its active state.

Thus, the option of Jean Lave for an analytical focus on direct experience in a lived-in-world, in a way induces

> reformulating the role of direct experience raising the question of how activity is made accountable while ongoing. An analytic focus on direct experience in the lived-in world leads to emphasis on a reflexive view of the constitution of goals in activity and the proposition that goals are constructed. (Lave 1988, p. 183)

This does not seem to be compatible with a linear view of action as directed towards established goals – "*action is not 'goal directed' nor are goals a condition for action*" (p. 183). Taking as support the idea from Wittgenstein and Giddens that it is through the recursive character of social life that it is possible to capture the nature of social practices as a continuous process of production and reproduction, Lave concludes that "*the meaning of activity is constructed in action*" (p. 184). Where the intentional character of activity comes from?

> In this perspective, motivation is neither merely internal to the person nor to be found exclusively in the environment. That is, even as goals are not 'needs' (hunger or sexual desire are socially constituted in the world), they are not prefabricated by the person-acting or some other goal-giver as a precondition for action. And activity and its values are generated simultaneously, given that action is constituted in circumstances which both impel and give meaning to it. Motivation for activity thus appears to be a complex phenomenon deriving from constitutive order in relation with experience. (Lave 1988, p. 184)

More than adding a typical approach from activity theory (for whom the external world is determinant) with a phenomenological reading (that gives the 'power' to individuals) there is a possibility to dialectically integrate aspects of the two theoretical fields that allow one to argue that setting and activity connect with the mind through its constitutive relations with person-acting (Santos and Matos 2000). Thus, instead of talking of goals (as in activity theory) a situated perspective refers to "*expectations, dialectically constituted in gap-closing processes, enable activity while they change in the course of activity backward and forward* ***in*** *time at the* ***same*** *time*" (Lave 1988, p. 185, emphasis in the original). This is closely related to the way Jean Lave conceptualizes intentions of actors in ongoing activity as they are "*engaged in what they are doing. When that activity poses conflicts, difficulties, in short dilemmas, they engage in resolving them*" (Lave 1992, p. 80). The procedures adopted in solving them gain form and meaning in relation to those dilemmas that are finally what motives their practices. It is the specific character of certain conflicts more adequate to concrete action that shapes what are problems to be solved. What makes a certain situation be seen as a dilemma would be what makes it be seen as a problem deserving effort in its resolution – what is seen as problematic in the activity emerges from and within that activity. This echoes the notion of contradiction (Engeström 1991) and its role in the activity that I will discuss later.

Mediating Artefacts in Activity Theory and Structuring Resources in Situated Learning

Resources are ways through which transformative relations are incorporated in the production and reproduction of social practices (Giddens 1996). This means that resources are intimately connected to power, be it seen either as an ability that transforms activity or adopting specific sense of domination or ability to intervene. Resources are always means through which social power is implemented; they are the basis and the vehicles of power. Given that resources are equally structural components of social systems, they become also the means through which the structures of domination are reproduced. It is within this framework that Giddens considers that exerting power is not a type of action; power is instantiated in action as regular and routine phenomenon. In this sense, power is not a resource but it depends on resources (Santos and Matos 2008).

A strong claim of the mediating role of artefacts seems to be clear in the introduction of the book edited by Dorothy Holland and Jean Lave in 2001 (Holland and Lave 2001). The authors assume a theoretical perspective grounded in a theory of practice that emphasizes the processes of social formation and cultural production and look with particular attention to cultural forms (close to the conceptualization of cultural artefact with its materiality) given the power of inscription they have. This echoes previous developments of Lave (1998) and Lave and Wenger (1991) on structuring resources that constitute authentic mediating artefacts.

This discussion has family resemblances to the notion of zone of proximal development (zpd). In an extremely clarifying analysis, Meira and Lerman (2009) criticized the way zpd is conceptualized as a field, a sort of physical space that children get in and that the adult (the teacher) is supposed to reach to be able to successfully teach the child. They present zpd as a symbolic space emerging from learning as a product of dialogic interaction. Thus zpd is thought of as a future oriented structure. Meira and Lerman (2009) conceptualize zpd as a tool to analyse teaching and learning environments at school and they consider a need to make further steps in conceptualizing zpd "from being thought of as a physical entity, towards the notion of a sign-mediated, intersubjective space for analysing how people become actors and communicators within any given activity or social practice" (p. 1). They argue that the revolutionary role of zpd is related to the idea of symbolic space where learning leads development and dialectic thinking and speech is manifested and where the individual's meanings encounter social meanings and purposes. "This implies that the opportunity and possibility for learning does not exist prior to the event or activity" (p. 1).

Participation in Social Practices

The concept of artefact and its relation to the idea of resource brings along the need to discuss the concept of participation which is central in situated learning perspectives. The concept of participation in social practices (within a situated view) and the idea of transformation in activity systems (from activity theory) must be addressed dialogically and unpacked. Sfard (2006) makes an insightful approach to the notion of participation juxtaposing a participationist discourse in mathematics learning against an acquisitionist metaphor claiming that the way we frame learning has powerful consequences both in research and in teaching practices.

It is important to note how researchers in education implicitly associate learning with the ability to participate, avoiding a positioning that puts learning as acquisition but bringing in the concepts such as apprenticeship, guided participation and scaffolding (Rogoff 1990). A number of researchers in mathematics education use, in a more or less explicit form, those ideas to study a variety of learning contexts and problems. However, it is not apparent that the notion of participation is assumed as embedding human agency in the social world in a constitutive form.

The concept of participation is present in all perspectives that claim a situated nature of learning (Santos 2004). It is in fact, as Santos argues, in that common understanding of the centrality of participation that most situated perspectives connect into learning. Lave and Wenger (1991) claim learning as situated in "*legitimate peripheral participation in communities of practice*" (p. 122); Greeno (1993) considers learning as a process of people becoming more capable of participating in practices; and Rogoff (1990) views participation as a process and a product, clamming that it is through guided participation (in systems of apprenticeship) that cognitive development occurs as participatory appropriation. Within those perspectives, the units of analysis include person, activity and the contexts where activity takes place.

Lave and Wenger (1991) identify two rather useful elements to characterize participation in order to reflect on learning:

(i) the social organization, where power relations shape the categorization and forms of participation of people, and
(ii) the relationships between participation within the activity and the recognition of the relevance that participation has to their life projects.

They widely explored the three dimensional concept of legitimate peripheral participation arguing it can be fruitfully expanded and introduced here for its potential to open space to articulate the idea of learning-in-practice with the notion of acting in an activity system. The three dimensions are as follows.

First, legitimacy of participation is a characteristic of participation that refers to the possibilities and degree of openness for action within the community on the part of the participant. The possibilities of participation are not exclusively dependent on rules and norms (both explicit and implicit), they articulate to the affordances and constraints offered by the community; it thus defines belonging not only as a crucial condition of learning but also as constitutive of learning. There are several (although equally legitimate) forms of belonging to a community, linked to more or less inclusive ways of being, located (by the collective) in the field of participation implicitly defined by the community. Participating in what is peculiar and essential in the practice (and not necessarily oriented towards its learning) confers legitimacy to participation. In fact participating is the legitimate way of accessing the practice and of being recognized as a participant. But the concept of legitimacy of participation opens four ways to dynamically conceptualize participation in a community (Lave and Wenger 1991): full participation (as an insider); full non-participation (as an outsider); 'inbound' participation (heading towards full participation); and 'outbound' participation (heading towards full non-participation). Those four spaces are to be understood as conceptual categories that do not categorize nor classify participants but that instead create horizontal landscapes allowing trajectories of participation. Mediation artefacts – such as language – may play a major role in the process of gaining legitimacy of participation. As Lerman (1994) states "*language is specific to particular social practices and is associated with power as knowledge and knowledge as power; language structures what we can talk about*" (p. 193).

Second, the peripherality of participation refers to the positioning the subject (or collective) takes in a certain practice; that is, it localizes the subject/participant in the activity systems where participation occurs; peripherality is related to the nature of the engagement of the participant and to its several forms; understanding the positioning of the person in the field of possibilities of participation opens ways to dynamically clarifying where the participants are heading and that's why it is associated to the idea of 'trajectory of participation'. Although it brings in a topological metaphor, peripherality of participation does not refer to a metric in relation to a standard form or degree of participation and therefore it is not opposed to the idea of 'central' participation (which has no meaning in a community of practice). The topological metaphorical space of participation has multiple dimensions turning 'central' a meaningless idea and in fact a misleading metaphor. Peripherality translates into multiple forms of participation and to the possibility of several and different forms of involvement; however, it is both participant and community who dynamically define peripherality whereby apparent changes in the positioning and perspective are seen as natural (e.g. typical patterns of the specific practice) both in terms of the trajectories of participation and in the development of participants' identity.

Third, legitimacy of peripherality is a notion implicated in social structures thus involving relations of power. In an activity system, power and associated mediation is constituted according to the legitimacy of participation and it is inherent to the trajectory of participation (and thus to the learning curriculum) (Wenger 1998). The issue of the legitimacy of peripherality and the mechanisms implicated in its development stand as conditions that allow participation. The concept of legitimacy of peripherality makes explicit an inherent ambiguity in participation: if peripherality is legitimated through the access to an increasing and more intense participation, the subject faces a position that progressively gives power to those who learn; if, on the contrary, participation does not develop (e.g. because there is legitimacy in avoiding a stronger engagement and participation) the subject faces a positioning that closes the access to a more powerful stance; the ambiguity in peripheral participation links to the matter of legitimacy, of social organization of resources and control over them (Lave and Wenger 1991).

Learning as Participatory Transformation

Within activity theory, learning can be addressed as an integral form of development that is materialized in qualitative transformations of the activity system, on a macro-level of analysis (e.g. within the social world where the students' practice unfolds) or/and of the subject, from a micro-analytical perspective (e.g. assuming the perspective of the student). This movement is mainly related to progression towards a wider and expansive field, for both the subject and the context. The need to bring into dialogue the analysis of collective activity systems and the point of view of individual subjects (Engeström and Sannino 2010) can be addressed through the exploration of the idea of learning as participatory transformation (Matos 2010).

Learning as transformation is inherently connected to the idea of learning as an activity, or more likely, as learning activity. Learning activity only gets sense and meaning when understood within a system (the activity system), which is representative of the established relationships between the subject and the social world. It is central to consider and analyze how such activity systems change and get transformed over time.

From the point of view of learning as transformation, knowledge is seen and considered as unstable, volatile, diffuse, emergent and in constant evolution. Assuming knowledge as existing in the relations of person to the artefacts, to the other (in the practice of the community) within the activity system, we should consider both the vertical and hierarchical processes of learning (which are not denied) as well as processes of "*horizontal and sideways learning and development*" (Engestrom 2001, p. 153), where the boundaries of knowledge are open, not imprisoned but crossed.

Learning is understood here as expansive transformation (Engeström 1991) as the activity systems move up through cycles of qualitative change, through which the motive of the activity is (re)conceptualized and new and radically broader horizons of possibilities are embraced.

In consonance with the perspective of Meira and Lerman (2009), a complete cycle of expansive transformation can be seen as a collective journey through the system's zone of proximal development conceptualized as:

> (. . .) the distance between the present everyday actions of the individuals and the historically new form of the societal activity that can be collectively generated as a solution to the double bind potentially embedded in the everyday actions. (Engeström 1991, p. 174)

Among the components of the activity (learning) system continuing and constant changes are happening. The activity system is incessantly rebuilding itself, and these internal reconstructions are seen as attempts of reorganization or (re)mediation of the system, that take place in order to resolve internal and external contradictions.

This perspective on learning – coined by Engeström as *learning as expansive transformation* – emerges as an historically more advanced view on learning. It reveals other dimensions of learning in connection with forms of participation showing the driving-forces that seem to be responsible for the processes through which humans transcend their given contexts.

Expansive learning puts into question the sense and meaning of the context and established norms that are questioned, leading to the emergence and construction of alternatives. An important implication is that learning itself produces culturally expansive new patterns and forms of activity. The changes in human activity (and by extension, in the organization where they take place) are considered instances of expansive transformation.

Taking learning as participatory transformation is configured in six characterizing issues (LEARN 2010) as follows.

Learning Develops in Collectives

Learning-in-practice is conceptualized as a collective endeavor taking place within activity which has and reveals a social nature. Human development is seen as arising from social interactions. Although this assumption does not deny individual learning, it goes beyond an individualistic perspective and integrates learning as acting for and by collective purposes. Consequently, human development is seen as resulting from collective transformation, which is historical and culturally contextualized and shared.

Learning Is a Contextual Phenomenon

In the analysis and understanding of learning it is essential to consider the socio-historical context in which it unfolds. The socio-historical dimension of human learning is central emphasis being put on the ecological character of learning. Context is considered not only the space-time frame that is directly incorporated in the activity, but also, in a wider perspective, the historic social political and economic time, where activity takes place. This is crucial to understand the learning-in-practice.

Learning Has a Dynamic Nature

Learning as (expansive) participatory transformation assumes a dynamic character through the creative and expanding movements resulting from the reconstruction of subjects – individual and/or collective. In an activity system, relationships between elements evolve leading to changes in the structure of the system. Central to the movements and processes of expansive transformation (as source of human learning and development) is the notion of contradiction (Engeström 1987). Internal and external contradictions constitute the driving forces of change in human activity as learning is linked to the dynamic resolution of emerging contradictions in activity. Contradictions do not show directly but appear as disturbances, disruptions, innovations and changes in activity systems. Transformation takes place by cyclical movements of resolving contradictions, which are typically associated with the development of activity (Engeström 1987).

Learning Is Necessarily Intentional

This is because it has an intentional basis. Behind any human activity there are always motives that drive, orient and maintain subjects' activity. The motives are connected to existing needs (sometimes expressed as desires) which are seen as being fulfilled acting over particular objects (material or imagined, but possibly explicit). Objects and needs by themselves do not produce activity, therefore motive should be necessarily integrated in the learning process.

Learning Is Intrinsically Linked to Production

This is because learning activity reveals always a productive nature. It's linked to the transformation of an object into a given outcome or result. The essence of learning activity is the production of new structures of social activity, which includes new objects, new tools, and new activities. The objects of human learning activity can be pointed out to be their own social productive practices. The object of learning activity is the societal productive practice (or the social world) in its full diversity and dynamic complexity (Engeström 2001). Thus, considering learning as transformation imposes the avoidance of seeing learning as reproduction, and adopting learning as creation and innovation.

To Conclude

Approaching mathematics learning from a situated point of view and locating learning-in-practice within an activity system, relates to the very idea of learning as participatory transformation. This reinforces three key ideas about the human social role:

(i) the person is a systemic, social and historically registered being;
(ii) the person is a creator and a transformer of collective subjectivity;
(iii) the person cannot be fully understood without its cultural means (artefacts).

Jean Lave (1996) presents three dimensions that every theory of learning should include:

(i) a learning telos, meaning a direction of changing and transformation,
(ii) learning mechanisms, as the ways that learning happens, and
(iii) the relation subject-social world as the general specification of relationships between subjects and social world (not as learners and learned stuff) which represents the key issue in social theories of learning.

I understand learning telos in relation to the issue of the relevance of the point of view of the subject (individual or collective) in transforming processes assuming that subjects (and collectives) are oriented towards recognition and identity and act in order to become participants in forms of distributed knowledge within the community. Learning is seen as transformation while subjects engage in goal oriented activity towards the transformation of objects that reify their needs-based motives and wills. Perhaps one could say as Wenger has (personal communication, 18 May 2010) that the telos of activity theory is more focused on the transformation of the object while that of participation is more focused on the transformation of the person.

(continued)

Learning mechanisms are conceptualized as the different forms of becoming a participant in social practices and includes the mechanisms of legitimate peripheral participation, engagement, alignment, imagination – components of ways of belonging, according to Wenger (1998), and the development of meaning. Learning mechanisms are seen as processes (that drive moves into change) such as internalization and externalization and mediation. The discussion in Lerman (2000b) when Steve asks "*to what extent, though, does Vygotsky's perspective provide the mechanism to which Lave refers?*" (p. 367) is extremely relevant and insightful. While Lave suggests that the need for learning mechanisms disappears into practice and people becoming kinds of persons, Steve stresses that "*becoming kinds of persons still calls for a mechanism*" (p. 368) and he proposes that "*internalization through semiotic mediation in the zpd is a suitable candidate*" (p. 368). At another level, to elaborate on the issue of the learning mechanisms we could look at them as actions that open learning possibilities (e.g. dialogue, reflection, intention, critique) and participation in dialogue, acting and producing meaning (Alrø and Skovsmose 2004). This seems to be a challenging topic that certainly deserves further study.

Addressing learning as transformation, as subject and community act upon an object (which goes from something potential into an outcome charged with new meanings and new forms of talk) it is crucial in this process that contradictions are identified (from conflicts and perturbations) and efforts are made to handle them (Engeström 2001). It is in the process of dealing and overcoming the contradictions that crucial action is taken and learning occurs.

If we take learning as participatory transformation the subject is viewed as agent in the socio-historical construction of the world and as product of that construction and culture. The subject is only understood in relation to cultural means (artefacts) of access to knowledge. This view puts value both on vertical and horizontal relationships. The way the relationships between person and social world are expressed assumes the subject as agent in a social world in conflict and thus inviting strategies for control and success. As to specify relationships between subjects, communities and the social world, person and world are not separable entities. Persons within their practices and the social world are mutually constitutive.

I hope that this contribution to articulate different dimensions of learning within different historically situated theories stimulates the emergence of interrogations about mathematics learning. I started this article with a quotation from Steve Lerman that reflects my positioning towards all issues in mathematics education research and practice. It certainly suggests relevant implications emerging from the discussion I presented in this article.

References

Abreu, G., Bishop, A., & Presmeg, N. (2002). *Transitions between contexts of mathematical practices*. Dordrecht: Kluwer.

Alrø, H., & Skovsmose, O. (2004). *Dialogue and learning in mathematics education: Intention, reflection, critique*. Dordrecht: Kluwer.

Bakhurst, D. (1997). Activity, consciousness, and communication. In M. Cole, Y. Engestrom, & O. Vasquez (Eds.), *Mind, culture and activity: Seminal papers from the Laboratory of Comparative Human Cognition* (pp. 147–163). Cambridge: Cambridge University Press.

Davydov, V. V. (1999). The content and unsolved problems of activity theory. In Y. Engestrom, R. Miettinen, & R.-L. Punamaki (Eds.), *Perspectives on activity theory* (pp. 39–52). Cambridge: Cambridge University Press.

Davydov, V. V., & Radzikhovskii, L. A. (1985). Vygotsky's theory and activity-oriented approach in psychology. In J. V. Wertsch (Ed.), *Culture, communications and cognition: Vygotskian perspectives*. Cambridge: Cambridge University Press.

Engeström, Y. (1987). *Learning by expanding: An activity theoretical approach to developmental research*. Helsinki: Orienta-Konsultit.

Engeström, Y. (1991). Non scloolae sed vitae discimus: Toward overcoming the encapsulation of school learning. *Learning and Instruction, 1*, 243–259.

Engestrom, Y. (1999). Innovative learning in work teams: Analysing cycles of knowledge creation in practice. In Y. Engestrom, R. Miettinen, & R.-L. Punamaki (Eds.), *Perspectives on activity theory* (pp. 377–404). Cambridge: Cambridge University Press.

Engeström, Y. (2001). Expansive learning at work: Toward an activity theoretical reconceptualization. *Journal of Education and Work, 14*(1), 133–156.

Engeström, Y. (2007). Enriching the theory of expansive learning: Lessons from journeys toward co-configuration. *Mind, Culture, and Activity, 14*(1–2), 23–29.

Engestrom, Y., & Cole, M. (1997). Situated cognition in search of an agenda. In D. Kirshner & J. A. Whitson (Eds.), *Situated cognition* (pp. 301–309). Mahwah: Lawrence Erlbaum Associates.

Engeström, Y., & Sannino, A. (2010). Studies of expansive learning: Foundations, findings and future challenges. *Educational Research Review, 5*, 1–24.

Even, R., & Ball, D. (2009). *The professional education and development of teachers of mathematics*. New York: Springer.

Giddens, A. (1996). *Novas regras do método sociológico* (António Escobar Pires, Trad.). (2ª ed., Vol. 34). Lisboa: Gradiva.

Greeno, J. (1993). For research to reform education and cognitive science. In L. A. Penner et al. (Eds.), *The challenge in mathematics and science education: Psychology response* (pp. 153–192). Washington, DC: American Psychology Association.

Hedegaard, M., Chaiklin, S., & Jensen, U. J. (1999). Activity theory and social practice: An introduction. In S. Chaiklin, M. Hedegaard, & U. J. Jensen (Eds.), *Activity theory and social practice* (pp. 12–30). Aarhus: Aarhus University Press.

Holland, D., & Lave, J. (2001). History in person: An Introduction. In D. Holland & J. Lave (Eds.), *History in person: Enduring struggles, contentious practice, intimate identities* (pp. 3–33). Santa Fe: School of American Research Press.

Jensen, U. J. (1999). Categories in activity theory: Marx's philosophy just-in-time. In S. Chaiklin, M. Hedegaard, & U. J. Jensen (Eds.), *Activity theory and social practice* (pp. 79–99). Aarhus: Aarhus University Press.

Kanes, C., & Lerman, S. (2008). Developing the concept of community of practice. In A. Watson & P. Winbourne (Eds.), *New directions for situated cognition in mathematics education* (pp. 310–326). New York: Springer.

Lave, J. (1988). *Cognition in practice: Mind, mathematics and culture in everyday life*. Cambridge, MA: Cambridge University Press.

Lave, J. (1992). Word problems: A microcosm of theories of learning. In P. Light & G. Butterworth (Eds.), *Context and cognition: Ways of learning and knowing* (pp. 74–92). Hemel Hempstead: Harvester Wheatsheaf.
Lave, J. (1996). Teaching, as learning, in practice. *Mind, Culture, and Activity, 3*(3), 149–164.
Lave, J. (1997). The culture of acquisition and the practice of understanding. In D. Kirshner & J. Whitson (Eds.), *Situated cognition* (pp. 17–35). London: LEA, Publishers.
Lave, J., & Wenger, E. (1991). *Situated Learning: Legitimate peripheral participation.* Cambridge: Cambridge University Press.
Lave, J., Murtaught, M., & de la Rocha, O. (1984). The dialectic of arithmetic in grocery shopping. In B. Rogoff & J. Lave (Eds.), *Everyday cognition: Its development in social context* (pp. 67–94). Cambridge: Harvard University Press.
LEARN, Project. (2010). *Learning framework.* Lisbon: University of Lisbon.
Leont'ev, A. N. (1978). *Activity, consciousness and personality* (Marie J. Hall, Trans.). Englewood Cliffs: Prentice-Hall.
Lerman, S. (1992). Learning mathematics as a revolutionary activity. In S. Lerman & M. Nickson (Eds.), *The social context of mathematics education: Theory and practice* (pp. 170–179). London: Southbank Press.
Lerman, S. (1994). Changing focus in the mathematics classroom. In S. Lerman (Ed.), *Cultural perspectives on the mathematics classroom* (pp. 191–213). Dordrecht: Kluwer.
Lerman, S. (1996). Intersubjectivity in mathematics learning: A challenge to the radical constructivist paradigm? *Journal for Research in Mathematics Education, 27*(2), 133–150.
Lerman, S. (1998). Learning as social practice: An appreciative critique. In A. Watson (Ed.), *Situated cognition and the learning of mathematics* (pp. 33–42). Oxford: Centre for Mathematics Education Research, University of Oxford Department of Educational Studies.
Lerman, S. (2000a). The social practices of the mathematics classroom. In J.-F. Matos & E. Fernandes (Eds.), *Investigação em educação matemática: Perspectivas e problemas* (pp. 357–370). Lisboa: Universidade de Madeira: Associação de Professores de Matemática.
Lerman, S. (2000b). The social turn in mathematics education research. In J. Boaler (Ed.), *Multiple perspectives on mathematics teaching and learning* (pp. 19–44). Westport: Ablex.
Lerman, S. (2001). Getting used to mathematics: Alternative ways of speaking about becoming mathematical. *Ways of Knowing, 1*(1), 47–52.
Lerman, S. (2006). Theories of mathematics education: Is plurality a problem? *Zentralblatt für Didaktik der Mathematik, 38*(1), 8–13. doi:10.1007/bf02655902.
Matos, J. F. (2010). Towards a learning framework in mathematics: Taking participation and transformation as key concepts. In M. Pinto & T. Kawasaki (Eds.), *Proceedings of the 34th conference of the International Group for the Psychology of Mathematics Education* (Vol. 1, pp. 41–59). Belo Horizonte: Universidade de Minas Gerais.
Matos, J. F., van Dormolen, J., Groves, S., & Zan, R. (2002). Learning from learners. In *Proceedings of the 26th international conference for the Psychology of Mathematics Education: PME* (Vol. 1, pp. 83–109). Norwich: University of East Anglia.
Matos, J. F., Powell, A., & Sztajn, P. (2009). Mathematics' teachers professional development: Processes of learning in and from practice. In R. Even & D. Ball (Eds.), *The professional education and development of teachers of mathematics* (pp. 167–184). New York: Springer.
Meira, L., & Lerman, S. (2009). Zones of proximal development as fields for communication and dialogue. In C. Lightfoot & M. C. D. P. Lyra (Eds.), *Challenges and strategies for studying human development in cultural contexts* (pp. 199–219). Rome: Firera Publishing.
Miettinen, R. (1999). Transcending traditional school learning: Teachers' work and networks of learning. In Y. Engestrom, R. Miettinen, & R.-L. Punamaki (Eds.), *Perspectives on activity theory* (pp. 325–344). Cambridge: Cambridge University Press.
Rogoff, B. (1990). *Apprenticeship in thinking: Cognitive development in social context.* New York: Oxford University Press.
Santos, M. (2004). *Encontros e Esperas com os Ardinas em Cabo Verde.* PhD dissertation, Universidade of Lisboa.

Santos, M., & Matos, J. F. (1998). School mathematics learning: Participation through appropriation of mathematical artefacts. In A. Watson (Ed.), *Situated cognition and the learning of mathematics* (pp. 104–125). Oxford: Centre for Mathematics Education Research.

Santos, M., & Matos, J. F. (2000). Thinking about mathematical learning with Cabo Verde 'ardinas'. In A. Ahmed, H. Williams, & J. Kraemer (Eds.), *Cultural diversity and mathematics education* (pp. 383–389). Chichester: Horwood Publishing.

Santos, M., & Matos, J. F. (2002). Thinking about mathematical learning with Cabo Verde ardinas. In G. Abreu, A. Bishop, & N. Presmeg (Eds.), *Transitions between contexts of mathematical practices* (pp. 81–122). Dordrecht: Kluwer.

Santos, M., & Matos, J. F. (2008). The role of artefacts in mathematical thinking: A situated learning perspective. In A. Watson & P. Winbourne (Eds.), *New directions for situated cognition in mathematics education* (pp. 179–204). New York: Springer.

Sfard, A. (2006). Participationist discourse in mathematical learning. In J. Maasz & W. Schloeglmann (Eds.), *New mathematics education research and practice* (pp. 153–170). Rotterdam: Sense Publishers.

Toulmin, S. (1999). Knowledge as shared procedures. In Y. Engestrom, R. Miettinen, & R.-L. Punamaki (Eds.), *Perspectives on activity theory* (pp. 53–64). Cambridge: Cambridge University Press.

Wenger, E. (1998). *Communities of practice: Learning, meaning and identity*. Cambridge: Cambridge University Press.

Chapter 14
The Philosophy of Mathematics Education: Stephen Lerman's Contributions

Paul Ernest

Introduction: What Is the Philosophy of Mathematics Education?

In order to look at Steve Lerman's contributions to the philosophy of mathematics education it is first desirable to briefly map out the terrain, for the question of what constitutes the philosophy of mathematics education is not without multiple answers and ambiguity.

There is a narrow sense that can be applied in interpreting the words 'philosophy' and 'mathematics education'. The philosophy of some area or activity can be understood as its aims or rationale. Mathematics education understood in its simplest and most concrete sense concerns the activity or practice of teaching mathematics. So the narrowest sense of 'philosophy of mathematics education' concerns the aims or rationale behind the practice of teaching mathematics. 'What is the purpose of teaching and learning mathematics?' is an important question, perhaps the most central to this area of inquiry because its answer determines *why* we engage in these practices and *what* we hope will be achieved. I have included learning within the question because learning is inseparable from teaching. Although they can be conceived of separately, in practice teaching presupposes one or more learners. Only in pathological situations can one have teaching without learning, although of course the converse does not hold. Informal learning is often self-directed and takes place without explicit teaching.

It is important to note that aims, goals, purposes, rationales, etc., for teaching mathematics do not exist in a vacuum. They belong to people, whether individuals or social groups (Ernest 1991). Indeed since the teaching of mathematics is a widespread and highly organised social activity, and even acknowledging the

P. Ernest (✉)
Graduate School of Education, University of Exeter, Exeter, UK
e-mail: p.ernest@ex.ac.uk

P. Gates, R. Jorgensen (Zevenbergen) (eds.), *Shifts in the Field of Mathematics Education*, DOI 10.1007/978-981-287-179-4_14

existence of divergent multiple aims and goals among different persons, ultimately these aims, goals, purposes, rationales, and so on, need to be related to social groups and society in general. Aims are expressions of values, and thus the educational and social values of society or some part of it are implicated in this enquiry. In addition, the aims discussed so far are for the teaching of mathematics, so the aims and values centrally concern mathematics and its role and purposes in education and society.

So already by considering the narrow meaning of philosophy of mathematics education, the issues of the teaching and learning of mathematics, the underlying aims and rationales of this activity, the roles of the teacher, learner, and mathematics in society and the underlying values of the relevant social groups are implicated. This resembles the issues arising from applying Schwab's (1961) four *'commonplaces of teaching'* to the mathematics curriculum. These are the subject (mathematics), the learner of mathematics, the mathematics teacher, and the milieu of teaching, including the relationship of mathematics teaching and learning, and its aims, to society in general.

Stephen Lerman and the Philosophy of Mathematics Education

So where does Steve Lerman's work fit into these issues? As early as 1983 Steve began exploring the relevance of recent advances in the philosophy of mathematics on the teaching and learning of mathematics. Lerman (1983) contrasts approaches based on Imre Lakatos' (1976) fallibilist philosophy of mathematics with those based on the traditional and formal Euclidean style (to use Lakatos' term). Drawing on the Cockcroft Report's (1982) endorsement of problem solving and investigational Steve contrasted problem-solving with knowledge centred teaching styles showing that these were underpinned by different philosophies of mathematics. At the time this was a radical insight.

Steve's endorsement of investigational and problem solving teaching in mathematics was not uncritical and in Lerman (1989a) he forcibly argues that ill-focussed investigational activity in the classroom is less effective than traditional knowledge centred teaching, while continuing to argue against the dangers of drill (Lerman 1988).

However Steve's largest contribution to the Philosophy of Mathematics Education was probably made in his important and original PhD thesis (Lerman 1986). This explored alternative views of the nature of mathematics and their possible influence on the teaching of mathematics both theoretically and empirically. The theoretical part was an extensive exploration of Lakatos' fallibilist philosophy, its broader philosophical and sociological parallels and a critique of the traditional absolutist Euclidean philosophies of mathematics. The empirical part was a study of mathematics teachers' beliefs and classroom practices. The empirical study was published with a necessarily condensed theoretical justification in Lerman (1990a).

Steve's PhD thesis (Lerman 1986) had a big impact on me and influenced my own subsequent book (Ernest 1991) which helped to raise the prominence of the philosophy of mathematics education as a subfield of mathematics education research. Indeed Steve carefully read and helped improve my unpublished draft and even suggested the better title 'The Philosophy of Mathematics Education' instead of my original 'Mathematics Teaching Philosophies'. Steve contributed to the founding of a Philosophy of Mathematics Education group with regular conference contributions and in 1990 a journal (The Philosophy of Mathematics Education Journal) which continues successfully to this day. Indeed Steve edited the second issue (Lerman 1990c).

Broader Views of the Philosophy of Mathematics Education

There is a broader sense that can be applied in interpreting the philosophy of mathematics education beyond just focusing on the aims, rationale and basis for teaching mathematics. Some of the expanded senses include:

1. Philosophy applied to or of mathematics education;
2. Philosophy of mathematics applied to mathematics education or to education in general;
3. Philosophy of education applied to mathematics education (Brown 1995);
4. The application of philosophical concepts or methods (such as conceptual analysis) in mathematics education theories, research or methodology.

Each of these possible 'applications' of philosophy to mathematics education represents a different focus, and might very well foreground different issues and problems. However, this analysis in terms of applications might be taken to suggest that there are always substantive bodies of knowledge and applicational activities connecting them involved, whereas philosophy, mathematics education and other domains of knowledge encompass processes of enquiry and practice, personal knowledge, and as well as published knowledge representations. They are not simply substantial entities in themselves, but complex relationships and interactions between persons, society, social structures, knowledge representations and communicative and other practices.

Philosophy is about systematic analysis and the critical examination of fundamental problems. It involves the exercise of the mind and intellect, including thought, enquiry, reasoning and its results: judgements, conclusions beliefs and knowledge. There are many ways in which such processes as well as the substantive theories, concepts and results of past enquiry can be applied to and within mathematics education.

Why does philosophy matter? Why does theory in general matter? Because it enables people to see beyond the official story about the world, society, economics, education, mathematics, teaching and learning. It provides thinking tools for questioning the status quo, for seeing that 'what is' is not 'what has to be'; to see that the boundaries between the possible and impossible are not always where we are told they are, and it enables us to imagine alternatives. Just as literature can allow us to stand in other people's shoes and see the world through their eyes, so too philosophy and theory can give people new 'pairs of glasses' through which to see the world anew.

At the very least, this suggests that the philosophy of mathematics education should attend not only to he aims and purposes of the teaching and learning of mathematics (the narrow sense) or even just the philosophy of mathematics and its implications for educational practice. It suggests that we should look more widely for philosophical and theoretical tools for understanding all aspects of the teaching and learning of mathematics and its milieu. At the very least we need to look to the philosophy of Schwab's (1961) other commonplaces of teaching: the learner, the teacher, and the milieu or society. So we also have the philosophy of learning (mathematics), the philosophy of teaching (mathematics) and the philosophy of the milieu or society (in the first instance with respect to mathematics and mathematics education) as further elements to examine, and then we must also consider the discipline of mathematics education as a knowledge field in itself.

Looking at each of these four commonplaces in turn, a number of questions can be posed as issues for the philosophy of mathematics education, understood broadly, to address, including the following.

What Is Mathematics?

What is mathematics, and how can its unique characteristics be accommodated in a philosophy? Can mathematics be accounted for both as a body of knowledge and a social domain of enquiry? Does this lead to tensions? What philosophies of mathematics have been developed? What features of mathematics do they pick out as significant? What is their impact on the teaching and learning of mathematics? What is the rationale for picking out certain elements of mathematics for schooling? How can (and should) mathematics be conceptualised and transformed for educational purposes? What values and goals are involved? Is mathematics value-laden or value-free? How do mathematicians work and create new mathematical knowledge? What are the methods, aesthetics and values of mathematicians? How does history of mathematics relate to the philosophy of mathematics? Is mathematics changing as new methods and information and communication technologies emerge?

This already has begun to pose questions relating to the next area of enquiry.

How Does Mathematics Relate to Society?

How does mathematics education relate to society? What are the aims of mathematics education (i.e., the aims of mathematics teaching)? Are these aims valid? Whose aims are they? For whom? Based on which values? Who gains and who loses? How do the social, cultural and historical contexts relate to mathematics, the aims of teaching, and the teaching and learning of mathematics? What values underpin different sets of aims? How does mathematics contribute to the overall goals of society and education? What is the role of the teaching and learning of mathematics in promoting or hindering social justice conceived in terms of gender, race, class, (dis)ability and critical citizenship? Is feminist and/or anti-racist mathematics education possible and what does it mean? How is mathematics viewed and perceived in society? What impact does this have on education? What is the relationship between mathematics and society? What functions does it perform? Which of these functions are intended and visible? Which functions are unintended or invisible? To what extent do mathematical metaphors (e.g., profit and loss balance sheet) permeate social thinking? What is their philosophical significance? To whom is mathematics accountable?

What Is Learning (Mathematics)?

What assumptions, possibly implicit, underpin views of learning mathematics? Are these assumptions valid? Which epistemologies and learning theories are assumed? How can the social context of learning be accommodated? What are the philosophical presuppositions of constructivist, social constructivist, sociocultural and other theories of learning mathematics? Do these theories have any impact on classroom practice? What elements of learning mathematics are valuable? How can they be and should they be assessed? What feedback loops do different forms of assessment create, impacting on the processes of teaching and learning of mathematics? How strong is the analogy between the assessment of the learning of mathematics and the warranting of mathematical knowledge? What is the role of the learner? What powers of the learner are or could be developed by learning mathematics? How does the identity of the learner change and develop through learning mathematics? Does learning mathematics impact on the whole person for good or for ill? How is the future mathematician and the future citizen formed through learning mathematics? How important are affective dimensions including attitudes, beliefs and values in learning mathematics? What is mathematical ability and how can it be fostered? Is mathematics accessible to all? How do cultural artefacts and technologies, including information and communication technologies, support shape and foster the learning of mathematics?

What Is Teaching (Mathematics)?

What theories and epistemologies underlie the teaching of mathematics? What assumptions, possibly implicit, do mathematics teaching approaches rest on? Are these assumptions valid? What means are adopted to achieve the aims of mathematics education? Are the ends and means consistent? Can we uncover and explore different ideologies of education and mathematics education and their impact on teaching mathematics? What methods, resources and techniques are, have been, and might be, used in the teaching of mathematics? What theories underpin the use of different information and communication technologies in teaching mathematics? What sets of values do these technologies bring with them, both intended and unintended? Is there a philosophy of technology that enables us to understand the mediating roles of tools between humans and the world? What is it to know mathematics in a way that fulfils the aims of teaching mathematics? How can the teaching and learning of mathematics be evaluated and assessed? What is the role of the teacher? What range of roles is possible in the intermediary relation of the teacher between mathematics and the learner? What are the ethical, social and epistemological boundaries to the actions of the teacher? What mathematical knowledge does the teacher need? What impact do the teacher's beliefs, attitudes and personal philosophies of mathematics have on practice? How should mathematics teachers be educated? What is the difference between educating, training and developing mathematics teachers? What is (or should be) the role of research in mathematics teaching and the education of mathematics teachers?

One further set of questions for the philosophy of mathematics education goes beyond Schwab's four commonplaces of teaching, which were primarily about the nature of the (mathematics) curriculum. This further set concerns the status of mathematics education as a field of knowledge and coming to know in it.

What Is the Status of Mathematics Education as Knowledge Field?

What is the basis of mathematics education as a field of knowledge? Is mathematics education a discipline, a field of enquiry, an interdisciplinary area, a domain of extra-disciplinary applications, or what? Is it a science, social science, art or humanity, or none or all of these? What is its relationship with other disciplines such as philosophy, mathematics, sociology, psychology, linguistics, anthropology, etc.? How do we come to know in mathematics education? What is the basis for knowledge claims in research in mathematics education? What research methods and methodologies are employed and what is their philosophical basis and status? How does the mathematics education research community judge knowledge claims? What standards are applied? What is the role and function of the researcher in mathematics education? What is the status of theories in mathematics education?

Do we appropriate theories and concepts from other disciplines or 'grow our own'? Which is better? How have modern developments in philosophy (phenomenology, critical theory, post-structuralism, post-modernism, Hermeneutics, semiotics, etc.) impacted on mathematics education? What is the impact of research in mathematics education on other disciplines? Can the philosophy of mathematics education have any impact on the practices of teaching and learning of mathematics, on research in mathematics education, or on other disciplines? What is the status of the philosophy of mathematics education itself?

These five sets of questions encompass, in my view, most of what is important for the philosophy of mathematics education to consider and explore. These sets are not wholly discrete, as various areas of overlap reveal. Some of the questions are not essentially philosophical, in that they can also be addressed and explored in ways that foreground other disciplinary perspectives, such as sociology and psychology. However, when such questions are approached philosophically, they become part of the business of the philosophy of mathematics education. And often to exclude certain questions *ab initio* is to adopt and promote a particular philosophical position, i.e., a particular philosophy of mathematics education.

These questions can be seen as 'bottom up' philosophical issues where it is the topics and problems addressed by mathematics education research that are used as a framework for charting the range of work in the philosophy of mathematics education. Of course admitting all of these questions to the philosophy of mathematics education greatly broadens the domain. It can then encompass much of the theoretical part of mathematics education altogether. In contrast, rather than looking at a mass of questions as in this 'bottom up' approach to defining the philosophy of mathematics education extensionally it is also possible to use a 'top down' approach. This takes the viewpoint of the different branches of philosophy as its framework for analysis. In my view it is easier to locate Steve Lerman's contribution to the philosophy of mathematics education by using such a 'top down' approach.

A 'Top Down' Analysis and Steve Lerman's Contributions

A 'top down' approach to the philosophy in mathematics education research uses the abstract branches of philosophy to provide the conceptual framework for analysis. Thus it considers research in mathematics education according to whether it draws on ontology and metaphysics, epistemology, social philosophy, ethics, methodology, aesthetics or other branches of philosophy.

Ontology and metaphysics have as yet been little applied in mathematics education research (but see Ernest 2012). Work drawing on aesthetics is still in its infancy (Ernest 2013; Sinclair 2008). However extensive uses of epistemology, social philosophy, ethics and methodology can be found in mathematics education research, and Steve Lerman has made significant contributions in each of these areas.

Epistemology

Epistemology concerns the theory of knowledge and can be taken to include both the nature of mathematical knowledge, including its means of verification, and the processes of coming to know, or learning. Thus the questions posed above (What is mathematics?) and (What is learning mathematics?) fall under this heading. Steve has continued his interest in exploring the relationships between epistemologies of mathematics and mathematics education discussed at the start of this chapter in Sierpinska and Lerman (1997). This provides a framework for examining some of the main epistemological questions concerning truth, meaning and certainty, and the different ways they can be interpreted for our field. It surveys a range of epistemologies including the contexts of justification and discovery, foundational and non-foundational perspectives on mathematics, critical, genetic, social and cultural epistemologies, and epistemologies of meaning. Looking within mathematics education a number of epistemological controversies are mapped out including the subjective-objective character of mathematical knowledge; the role in cognition of social and cultural context; relations between language and knowledge; and tensions between the major tenets of constructivism, socio-cultural views, interactionism and the French Didactique. Relationships between epistemology and a theory of instruction, especially in regard to didactic principles, are also considered, thus addressing the question 'What is teaching in mathematics?'.

Following his 'social turn' Steve applied Bernstein's (1999) sociological theories to mathematics and mathematics education. First of all, Steve made the contrast between the 'strong grammar' of mathematics and the sciences making them fields capable of generating unambiguous testable implications, with the 'weak grammar' of education and other humanities not so capable. Second, the use of Bernstein's distinction between vertical and horizontal discourse provided new insights into the structure of mathematics as made up of many adjacent fields of inquiry (Lerman 2010). These applications provide insights into the epistemology of mathematics, understood broadly, and by no means exhaust Steve's use of Bernsteinian theory.

Many of Steve's writings concern theories of learning. Early publications explored radical constructivism in mathematics education (Lerman 1989b). However like many of us he became aware of the problems arising from the individualistic nature of radical constructivist learning theory (Ernest 1994; Lerman 1996) and developed more socially orientated theories of learning (Lerman and Cowley 2010). His commitment to sociocultural theories of learning and his exploration of their philosophical underpinnings has seen in his writings for almost 20 years.

Social Philosophy

Steve Lerman is known for his use of the phrase 'the social turn' with respect to mathematics education research (Lerman 2000). This signifies his exploration, endorsement and leadership in developing and promoting social philosophy, social learning theory and sociological theory in mathematics education. Although his early publications already drew on social philosophers including Bloor and Wittgenstein by the mid-1990s Steve had adopted a strongly social view of mathematics and its learning and teaching. Lerman (1996) offers a powerful critique of radical and social constructivist of learning theories for mathematics from a sociocultural perspective. He went on to argue that many of the questions and problems in mathematics education research can be approached from socio-cultural foundations (Lerman 1998a, 2006). Indeed he has recently gone on to argue that the social construct of 'identity' can be used as a unit of analysis in researching and teaching mathematics (Lerman 2012).

Ethics

Ethics enters into mathematics education research in a number of ways including a concern with values, with social justice and equity approaches, and through the ethics of research methodology. Steve has contributed in each of these areas. In Lerman (1990b) he argues that despite its traditional value-free absolutist image mathematics is value laden. In Lerman (1992) he draws on Paulo Freire's emancipatory philosophy to argue that learning mathematics can be a revolutionary activity in the struggle for social justice and equity. Thirdly, he has jointly explored the ethics of different research methodologies in Adler and Lerman (2003). These are just sample publications because a concern with ethics and social justice is a strand that runs through many more of Steve's publications.

Methodology

Steve's publications and papers reflect an awareness of the philosophical issues underpinning research methods and methodologies in mathematics education. In particular he has explored the coordination and complementarities of different levels and units of analysis and coined the term 'zoom of a lens' (Lerman 1998b), also noted in other chapters in this volume. As this shows, one of his repeated concerns has been the status of theory in mathematics education and whether plurality is a problem and a single overarching theory is needed for consistency (Lerman 2010). Steve has also studied the different frameworks and philosophies underpinning research styles in mathematics education (Yore and Lerman 2008; Lerman et al. 2003a), as well as the philosophical stances of mathematics education researchers, be they critical public intellectuals or functional academics (Lerman et al 2003b).

Conclusion
In this chapter I have attempted to outline the sub-field of study the philosophy of mathematics education. I have characterized this in both narrow and broad terms, and from both bottom-up and top-down perspectives. One can characterize the area in terms of questions, and I have asked: What is the purpose of teaching and learning mathematics? What is mathematics? How does mathematics relate to society? What is learning mathematics? What is mathematics teaching? What is the status of mathematics education as knowledge field? Steve's work helps to address all of these questions. Additionally, I have characterized the sub-field of the philosophy of mathematics education using a 'top down' perspective using branches of philosophy. Looking at Steve's contributions to the epistemology, social philosophy, ethics, and methodology of mathematics education reveals both how extensive and deep his contributions are to mathematics education, to the theoretical foundations of our field of study, and to the philosophy of mathematics education. Without his contributions it would be a much impoverished field of study.

References

Adler, J., & Lerman, S. (2003). Getting the description right: Ethical practice in mathematics education research. In A. J. Bishop, M. A. Clements, C. Keitel, J. Kilpatrick, & F. K. S. Leung (Eds.), *Second international handbook on mathematics education* (pp. 441–470). Dordrecht: Kluwer.

Bernstein, B. (1999). Vertical and horizontal discourse: An essay. *British Journal of Sociology of Education, 20*(2), 157–173.

Brown, S. I. (1995, May). Philosophy of mathematics education: POM(E), PO(ME) OR POE(M)? *Philosophy of Mathematics Education Journal, 8*. Retrieved March 8, 2014, from http://people.exeter.ac.uk/PErnest/pome/pome_8.htm

Cockcroft Report. (1982). *Mathematics counts* (Report of the committee of inquiry on the teaching of mathematics). London: Her Majesty's Stationery Office.

Ernest, P. (Ed.). (1990–). *The Philosophy of Mathematics Education Journal*, Issues no. 1–27. Retrieved March 10, 2014 from http://people.exeter.ac.uk/PErnest/

Ernest, P. (1991). *The philosophy of mathematics education*. London: The Falmer Press.

Ernest, P. (1994). Social constructivism and the psychology of mathematics education. In P. Ernest (Ed.), *Constructing mathematical knowledge: Epistemology and mathematics education* (pp. 62–72). London: Falmer Press.

Ernest, P. (2012). What is our first philosophy in mathematics education. *For the Learning of Mathematics, 32*(3), 8–14.

Ernest, P. (2013, September 17–19). *Values and mathematics: Overt and covert*. Paper presented at *Mathematical Cultures* 2, London.

Lakatos, I. (1976). Proofs and refutations: The logic of mathematical discovery (J. Worrall & E. Zahar, Ed.). Cambridge: Cambridge University Press (To appear in *Culture and Dialogue, 4* (1), 2014 (Special issue, ed. M. Ovens)).

Lerman, S. (1983). Problem-solving or knowledge centred: The influence of philosophy on mathematics teaching. *International Journal of Mathematical Education in Science and Technology, 14*(1), 59–66.

Lerman, S. (1986). *Alternative views of the nature of mathematics and their possible influence on the teaching of mathematics*. Unpublished PhD thesis, King's College London.

Lerman, S. (1988). The dangers of drill ('Soundoff' article). *Mathematics Teacher, 81*(6), 412–413.
Lerman, S. (1989a). Investigations, where to now? In P. Ernest (Ed.), *Mathematics teaching: The state of the art* (pp. 73–80). Lewes: Falmer Press.
Lerman, S. (1989b). Constructivism, mathematics and mathematics education. *Educational Studies in Mathematics, 20*(2), 211–223.
Lerman, S. (1990a). Alternative perspectives of the nature of mathematics and their influence on the teaching of mathematics. *British Educational Research Journal, 16*(1), 53–61.
Lerman, S. (1990b). What has mathematics got to do with values? *The Humanistic Mathematics Network Newsletter, 5*, 29–33.
Lerman, S. (Ed.). (1990c). *The Philosophy of Mathematics Education Journal, 2*. Exeter: The University. Retrieved March 8, 2014, from http://people.exeter.ac.uk/PErnest/pome/pome2.htm
Lerman, S. (1992). Learning mathematics as a revolutionary activity. In S. Lerman & M. Nickson (Eds.), *The social context of mathematics education: Theory and practice* (pp. 170–179). London: Southbank Press.
Lerman, S. (1996). Intersubjectivity in mathematics learning: A challenge to the radical constructivist paradigm? *Journal for Research in Mathematics Education, 27*(2), 133–150.
Lerman, S. (1998a). Research on sociocultural approaches to mathematics education. In A. Sierpinska & J. Kilpatrick (Eds.), *Mathematics education as a research domain: A search for identity* (pp. 333–350). Dordrecht: Kluwer.
Lerman, S. (1998b). A moment in the zoom of a lens: Towards a discursive psychology of mathematics teaching and learning. In A. Olivier & K. Newstead (Eds.), *Proceedings of the twenty-second annual meeting of the International Group for the Psychology of Mathematics Education* (Vol. 1, pp. 66–81). Stellenbosch: University of Stellenbosch.
Lerman, S. (2000). The social turn in mathematics education research. In J. Boaler (Ed.), *Multiple perspectives on mathematics teaching and learning* (pp. 19–44). Westport: Ablex.
Lerman, S. (2006). Socio-cultural research in PME: Perspectives, trends and results. In A. Gutiérrez & P. Boero (Eds.), *Handbook of research on the psychology of mathematics education* (pp. 347–366). Rotterdam: Sense Publishers.
Lerman, S. (2010). Theories of mathematics education: Is plurality a problem? In B. Sriraman & L. English (Eds.), *Theories of mathematics education* (pp. 99–110). New York: Springer.
Lerman, S. (2012). 'Identity' as a unit of analysis in researching and teaching mathematics. In H. Daniels (Ed.), *Vygotsky and sociology* (pp. 175–191). London: Routledge.
Lerman, S., & Cowley, R. (2010). Learning theories in the mathematics classroom. In S. Johnston-Wilder & C. Lee (Eds.), *Leading practice and managing change in the mathematics department* (pp. 33–46). London: Tarquin.
Lerman, S., Xu, G.-R., & Tsatsaroni, A. (2003a). Developing theories of mathematics education research: The ESM story. *Educational Studies in Mathematics, 51*(1–2), 23–40.
Lerman, S., Xu, G.-R., & Tsatsaroni, A. (2003b). A sociological description of changes in the intellectual field of mathematics education research: Implications for the identities of academics. *Proceedings of the British Society for Research in Learning Mathematics, 23*(2), 43–48.
Schwab, J. J. (1961). Education and the structure of the disciplines. In I. Westbury, and N. J. Wilkof (Eds.), *Science, curriculum and liberal education: Selected essays of Joseph J. Schwab* (pp. 229–272). Chicago: University of Chicago Press, 1978.
Sierpinska, A., & Lerman, S. (1997). Epistemologies of mathematics and of mathematics education. In A. J. Bishop (Ed.), *International handbook of mathematics education* (Pt. 2, pp. 827–876). Dordrecht: Kluwer.
Sinclair, N. (2008, March 5–8). *Notes on the aesthetic dimension of mathematics education*. Paper presented at Conference on first 100 years of ICMI, Rome. Retrieved March 10, 2014, from http://www.unige.ch/math/EnsMath/Rome2008/ALL/Papers/SINCL.pdf
Yore, L., & Lerman, S. (2008). Metasyntheses of qualitative research studies in science and mathematics education. *International Journal of Science and Mathematics Education, 6*(2), 217–223.

Chapter 15
Lerman's Perspectives on Information and Communication Technology

Marcelo C. Borba and Ricardo Scucuglia

Introduction

In many of his recent publications, Steve Lerman discusses the use of information and communication technology (ICT) in mathematics teaching and learning, addressing several issues such as numeracy, classroom interactions, scaffolding, teacher education, pedagogy, online education, the use of whiteboards, and so forth (Lerman 2004; Lerman and Zevenbergen 2006, 2007; Zevenbergen and Lerman 2005, 2006, 2007, 2008; Crisan et al. 2006; Rosa and Lerman 2011).

On the one hand, within our work in mathematics education (e.g., Borba and Villarreal 2005; Borba 2012; Scucuglia 2012), we have addressed sociocultural perspectives to conceptualize the role of ICT, or digital technologies as we have been calling it lately, as cultural artefacts in mathematical learning and activity (Borba et al. 2010). We have built on the very notion of humans-with-media to emphasize cognition and mathematical knowledge production as a social, collective, and object-directed undertaking (Borba 2009). On the other hand, we have not properly deeply addressed Steve Lerman's perspectives in our theorization as we should. Thus, in this chapter, we present (a) the way Steve Lerman dealt with ICT in different publications and (b) potential links between his perspectives and part of the work of our research group on computers, other media and mathematics education – GPIMEM (http://www.rc.unesp.br/gpimem/) at Sao Paulo State University, Campus of Rio Claro, in Brazil. We also emphasize potential theoretical insights to our current interest on the use of digital technology and the performance arts for multimodal mathematical communication (Scucuglia 2012).

M.C. Borba (✉) • R. Scucuglia
State University of São Paulo (UNESP), São Paulo, Brazil
e-mail: mborba@rb.unesp.br; ricardos@ibilce.unesp.br

P. Gates, R. Jorgensen (Zevenbergen) (eds.), *Shifts in the Field of Mathematics Education*, DOI 10.1007/978-981-287-179-4_15

Sociocultural Perspectives for the Use of ICT as the Resistance Towards Constructivism

We find an interesting similarity on the use of sociocultural perspectives in theorizing pedagogies for the use of ICT in mathematics education when we compare some of Steve Lerman's ideas and the notion of humans-with-media proposed by Borba and Villarreal (2005). This notion has been paramount for most of research developed by GPIMEM. In order to emphasize a social, cultural, and collective nature of the use of ICT in education, authors such as Lerman, and Borba and Villarreal have pointed out some aspects that expose a kind of *resistance* toward the localism of the individual-biological nature of the being or subject in constructivist points of view. According to Lerman (1996):

> Rejecting constructivism, the individual is integrally part of the social world, and thinking is a dialect relationship with that world. Individual mental structures are not the fundamental unit of cognition; meanings, which are first on the social plane, perform this function. Inevitable biological development is not seen to lead to human functioning; the development of consciousness, which only takes place in social life, is the essence. (Lerman 1996, p. 148)

Steve Lerman also creates an argument to point out a tension within constructivist perspectives in terms of intersubjectivity, mainly through the movement involving radical and social constructivism.

> The extension of radical constructivism toward a social constructivism, in an attempt to incorporate intersubjectivity, leads to an incoherent theory of learning. A comparison of Piaget's positioning of the individual in relation to social life with that of Vygotsky and his followers is offered, in support of the claim that radical constructivism does not offer enough as an explanation of children's learning of mathematics . . . Constructivists, whether radical, weak, or social, draw their inspiration from Piaget, for whom the individual is the central element in meaning-making. . . . Vygotsky attempted to develop a fully cultural psychology by which I mean placing communication and social life at the center of meaning-making, which is a challenge to Piaget's ideas. (p. 133)

It is not a surprise to argue that sociocultural perspectives point out that reality, knowledge, and meaning are socially, historically, and culturally produced through language. Socioculturalism connects activity to participation in cultural practices (Cobb 1994). Instead of focusing on the individual processes of learners' meaning-making and knowledge construction (e.g., cognitive conflict and equilibrium in Piagetian constructivism), sociocultural perspectives emphasize the social interaction and enculturation of subjects in (mathematical) learning, development, and activity.

Vygotsky (1978) investigated children's development and learning and how these processes are conditioned by the role of culture and language. According to Vygotsky, higher mental functions are historically developed within particular cultural groups, through social interactions with the significant people in children's lives, particularly parents and teachers. Through these interactions, children learn the habits of the culture, including patterns of speech, verbal and written language, and other symbolic representations. Thus, Vygotsky emphasized (a) the social interaction with more knowledgeable others in the zone of proximal development and (b) the role of culturally developed sign systems and languages as psychological tools of thinking.

In fact, we do agree with Gadanidis and Geiger (2010) when they state that:

> Sociocultural theories of learning are founded on a position that intellectual development originates in, and so is not just facilitated by, social interaction. Learning is a process of enculturation into the practices of a learning community. Enculturation into the community requires the appropriation of modes of reasoning, discourse and knowledge creation that are accepted by the discipline around which the community is based. Learning mathematics in such a community means a learner must participate in debate about new ideas and practices, offer critique of others' ideas and defend their own propositions via explanations and justifications. (p. 96)

Central in our research, sociocultural theories actually lead us to an object-oriented view of cognition. Goos et al. (2000) clarify that "a central claim of sociocultural theory is that human action is mediated by cultural tools and is fundamentally transformed in the process" (p. 306). In our perspectives, technologies can be conceptualized as cultural artefacts of thinking (Papert 1980; Noss and Hoyles 1996). Borba and Villarreal (2005) thus argue that technologies are not neutral in mathematical knowledge production. Media are actors that (re)organize mathematical thinking. Not only humans, but humans-with-media (e.g., students-and-teachers-with-computers) produce mathematical knowledge.

Humans-with-Media

Borba and Villarreal (2005) use the expression *humans-with-media* as a metaphor to theorize the cognitive "inter-shaping" between humans and technologies in mathematical knowledge production. The inter-shaping relationship stresses the mutual shaping that there is between humans and artefacts. Artefacts are produced by collectives of humans-with-other-artefacts with a certain goal. Such a goal is transformed by others who use and shape it to social perspectives of a historically dated collective of humans-with-media. So artefacts and in particular digital artefacts are transformed and transform different collectives of humans-with-media, in the sense that artifacts are always seen as communicating device.

The authors build their perspectives using the notion of *technologies of intelligence* (Levy 1993): a historical-cognitive perspective of technologies. According to Levy (1993), there are three main technologies of intelligence associated with memory and knowledge. They are: orality, writing, and information technology. In oral societies, humans produced knowledge through myths and rituals, cyclically and locally, transmitting information from one generation to another. However, this *circularity* was reorganized into *linear* ways of reasoning in writing societies, mainly through the popularization of books, due in large part to the invention of Gutenberg's printer press.

ICT can be understood in the same way. The linearity of memory conditioned by the temporality of writing has been assuming a "web or net design" through the plasticity of digital technology. Computers and online tools combine multiple modes of communication. They shape the ways that contemporary societies interact and communicate. The "linear reasoning" of writing has been challenged by ways of thinking involving orality, writing, images, simulation, experimentation, and instantaneous communication. Regarding current technological innovations, there are innovating ways to communicate, extend memory, store information, represent, simulate, and produce meanings and knowledge.

Borba and Villarreal (2005) argue that "our individual consciousness and cognitive process are always subject to interaction with the technologies of intelligence" (p. 26). That is, "knowledge is produced with a given medium or technology of intelligence" (p. 23).

> Humans-with-media, humans-media or humans-with-technologies are metaphors that can lead to insights regarding how the production of knowledge itself takes place. . . . This metaphor synthesizes a view of cognition and of the history of technology that makes it possible to analyze the participation of new information technology 'actors' in these thinking collectives. (Borba and Villarreal 2005, p. 23)

Borba and Villarreal (2005) discuss sociocultural perspectives (Tikhomirov 1981) to develop the notion of humans-with-media. According to Tikhomirov (1981), computers do not replace, substitute, or merely complement humans in their intellectual activities. Processes mediated by computers *reorganize* thinking. Tikhomirov, who was Luria's student, argues that computers play a mediating role in thinking as language does in Vygotsky's theory. Regarding the nature of human-computer interaction in terms of feedback, the dimensions involving computational mediation provide new insights in terms of learning, development, and knowledge production. Tikhomirov claims that:

> With regard to the problem of regulation we can say that not only is the computer a new means of mediation of human activity but the very reorganization of this activity is different from that found under conditions in which the means described by Vygotsky are used. (p. 273)

Borba and Villarreal (2005) use Tikhomirov's ideas to argue how the notion of mediation by computers is qualitatively different to the mediation involving paper and pencil, for instance. Through digital mediation, information technologies reorganize mathematical thinking. Media shape knowledge production and transform mathematics.

Levy (1993) defines *cognitive ecology* as "the study of technical and collective dimensions of cognition" (p. 137). He sees technology not simply as a tool used by humans, but rather as an integral component of the cognitive ecology. Further Levy (1998) claims "as humans we never think alone or without tools. Institutions, languages, sign systems, technologies of communication, representation, and recording all form our cognitive activities in a profound manner" (p. 121). According to Levy, technologies *do not determine* thinking. Technologies *condition* thinking (Levy 1993, 2000). He (1993) uses the term *thinking collectives* to discuss the collaboration between human and non-human actors in the cognitive ecology. Levy (1993) argues that thinking collectives of humans-technologies form the cognitive ecology.

Levy (1997) relates cognitive ecology and thinking collectives to *collective intelligence,* defined as "a form of universally distributed intelligence, constantly enhanced, coordinated in real time, and resulting in the effective mobilization of skills" (p. 13). By intelligence, Levy (1998) means "the canonical set of cognitive aptitudes, namely the ability to perceive, remember, learn, imagine, and reason" (p. 123). More recently, Borba (2009, 2012) has proposed that media does not only change the way collectives think but it has changed the very nature of what "being human" means. Media such as mobile phone and computers are not only merging among themselves but are deeply transforming our very perception of who we-are-with-technology.

In fact, it is important to clarify that the resistance toward constructivist views, in the context of supporting a collective-sociocultural perspective for the educational use of ICT, does not exclude the symbiosis involving contextual and *personal* dimensions of classroom activities. In this direction, Crisan et al. (2007) propose a framework to theorize teachers' practices on use of ICT at the secondary level considering the involvement between ICT content and curricular conceptions, pedagogic and mathematical conceptions. The personal ICT pedagogical construct emerges from that involvement through teachers' learning and experiences with ICT (see Fig. 15.1). In fact, the authors argue that:

> Learning to teach with ICT is a process. It demands doing and practice and ... teachers developed their own 'expertise' with ICT, which we call here personal ICT pedagogical construct, consisting of conceptions of how the ICT tools and resources at their disposal

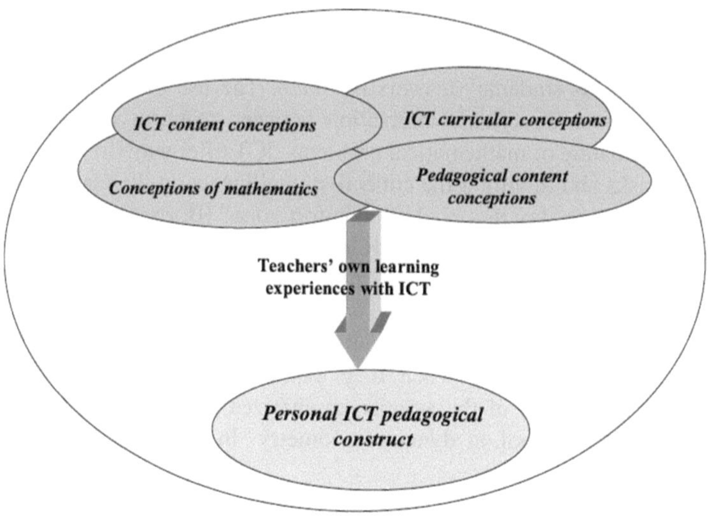

Fig. 15.1 Crisan, Lerman, and Winbourne's (2007) framework about teachers' practices on use of ICT

> benefited their teaching of mathematics and their pupils' understanding and learning of mathematics (Crisan et al. 2007, p. 33).

On the Risk Zone for Teaching with ICT

Lerman (2004) presents some aspects of a research project that focused on key policy initiatives in Queensland, with emphasis on numeracy and the use of ICT in the curriculum. In his conclusions, Lerman (2004) points out some problems toward *innovation* in teaching, highlighting that "most teachers are worried about being seen to be less competent in computer use than their students and fear the loss of power and status if students see they know more than their teachers" (p. 622). Similarly, Lerman and Zevenbergen (2006) argue that "there may be some resistance to change pedagogy in mathematics classrooms in response to the potential of ICTs and to the call for improving achievement amongst traditionally failing students" (p. 49).

Focusing specifically on some results of an analysis of the project called New Basics, Lerman and Zevenbergen (2006) state that:

> Of course there are many ways of using ICTs and not all of them enhance the learning of mathematics in the same way . . . [but some classrooms and pedagogies] may in fact offer the opportunity for successful learning by more students. We conjecture, however, that, without explicit awareness by teachers of the implications of different forms of pedagogy on different social groups the aims of the New Basics in terms of more equitable outcomes are not likely to be met. (p. 55)

In regard to these issues, Borba and Pentado (2001) suggest that the use of computers in education brings teachers to a "risk zone", because the use of ICT challenges the typical lesson structures in which teachers could predict and control every single event of the dynamics of the classroom such as the nature of the questions as well as students' answers for them. The use of ICT challenges the authoritative and unidirectional interaction between teachers and students and reorganizes the nature of mathematical problems. ICT offer ways to both to explore open-ended tasks and highlight the collective intelligence in the classroom. However, teachers tend to stay in a kind of "comfort zone" (Borba and Pentado 2001), resisting or ignoring the presence of ICT in the world or simply conducting a "domesticated" use of ICT, reproducing typical pedagogies of right or wrong, yes for no for the control of the classrooms.

Borba and Zulatto (2010) present examples of how mathematics teachers collaborate in an online course when they explore activities with the software Geometricks. The design of the activity emphasizes collective experimentation-with-technology and proof in dynamic geometry. In this scenario, the authors believe that:

> Teaching in online environments situates the teacher within a new model of risk zone with respect to the use of ICT in the teaching of mathematics. New challenges arise: How to follow the progress of my students who are physically distant? How to discuss mathematics

> online? How to express my reasoning? What resource is more appropriate for each situation? (Borba and Zulatto 2010, p. 114)

Borba and Zulatto (2010) also propose that there is also a dynamic of the risk zone, in which there may be teachers who actually feel comfortable running risks as they explore technology in the classroom. Teaching with technology "in online environments require teachers who are more comfortable working in the risk zone while learning together with their students/peers! Like engaging in 'radical sports,' with practice, the risk zone can become comfortable" (p. 124).

On the one hand, we do acknowledge the variety and diversity of pedagogies toward the use of ICT in mathematics education, the problem of accessibility of computers in education (mainly in the global south), and issues concerning technical support for teachers in schools as well as teachers' "computational literacy." On the other hand, the reorganization of thinking emergent with the use of ICT in mathematics (education) cannot be ignored in terms of cognition and affectivity, although the research conducted by GPIMEM has not properly addressed discussions on affectivity and the use of technology. Among several aspects, the use of ICT in education disrupts the power relations that see the teacher as the iconic symbol of knowledge in the classroom. As Doll (1993) points out, when the linear and sequential pedagogic dynamics become less ordered and more fuzzy, "the relations between teachers and students… change drastically" (p. 3). That is, "these relations … exemplify less the knowing teacher informing unknowing students, and more a group of individuals interacting together in the mutual exploration of relevant issues" (p. 4), and it has a direct influence in terms of curriculum.

On the Use of Interactive Whiteboards

Lerman and Zevenbergen (2007) mention how the affordances of interactive whiteboards (IWB) may offer possibilities for "rich communications and interactions in the classroom as teachers are seduced by the IWB's ability to capture pupils' attention" (p. 175). The authors highlight that:

> Teachers' advance preparation for using the IWB, often via the ubiquitous PowerPoint package or pre-prepared lessons for the IWB, are leading to a decreased likelihood that teachers will deviate in response to pupils' needs and indeed might notice pupils' needs less frequently through the possibility to increase the pacing of mathematics lessons. (p. 175)

Zevenbergen and Lerman (2007, 2008) explore teachers' use of IWB in classrooms through the various lenses of activity theory. These lenses help the authors to "understand the tensions and contradictions in teachers' use of the IWB and to identify possible developmental trajectories for realising some of their potential to change pedagogy for the better" (Zevenbergen and Lerman 2008, p. 124). The authors thus address the synergy between pedagogy and the use of ICT, seeing the classrooms as fruitful social environments, focusing the nature of the interaction of students-teachers-ICT in classrooms.

> The potential and rhetoric of IWB supporters, the ways in which it is used in the classroom may inhibit learning... The two dimensions that focus on knowledge production – intellectual quality and relevance – suggest that the scaffolding around the use of IWBs can be enhanced through higher expectations of learning... These aspects of pedagogy may be one way in which higher levels of intellectual quality may be facilitated. Aspects of the social environment – supportive school environment and recognition of difference – may also be challenged. The whole class interaction may stifle participation (and engagement) of students. Reorganising pedagogy so as to foster interaction, collaboration in smaller groups, or to employ other tools alongside the IWB may encourage greater interaction among learners. (Zevenbergen and Lerman 2008, p. 124)

The members of the research group GPIMEM did not conduct a research about the use of IWB in classrooms yet. However, Mazzi et al. (2012) conducted an exploratory study toward potential affordances of IWB for teaching and learning of Calculus and Geometry and produced a guide in Portuguese for a math-oriented use of a specific type of IWB (a guide is available at http://tidia-ae.rc.unesp.br/portal).

Exploring only the applications offered that IWB, Mazzi et al. (2012) identified some limitations such as (a) small dimension of the actual board interaction; (b) restrictions in transferring videos directly from websites; (c) imprecision of measurement tools and (d) higher costs of the IWB in Brazil. However, the authors highlighted the support of the IWB in running dynamic geometry software and CAS, that is, how typical software can be used with IWB. Thus, as mentioned by Zevenbergen and Lerman (2008), all those pedagogic issues regarding the use of ICT in mathematics education (see Tall 1991; Borba 1993; Noss and Hoyles 1996; Laborde 2000; Borba and Villarreal 2005) are also important issues toward the use of IWB.

We also see an interesting aspect of IWB in terms of multimodality (The New London Group 1996). The traditional modes for human-computer interactions happen through the use of screen, keyboards, mouse, speakers, microphone, webcams, and so on. In an IWB, one interacts directly by touching the screen instead of using a mouse. Thus, we do see a change from clicking to touching in terms of multimodality when we use an IWB, and when we use tablets. Since experimentation and visualization are fundamental aspects of mathematical exploration and thinking, we do conjecture that hands-on manipulation of virtual mathematical objects in an IWB has an impact in terms of heuristics and cognition. The transformation from clicking to touching is being properly addressed through the notion of humans-with-media in a current research project conducted by members of GPIMEM (Mazzi et al. 2012).

Humans-with-Internet: Performance and Identities

The use of the Internet in mathematics education (Borba 2004, 2009) and, more recently the use of performance arts and digital technology (Scucuglia 2012) have become an important research focus of our group GPIMEM. At this point in the chapter, we would like to highlight some conceptions we hold toward ICT in mathematics education:

- ICT has reorganized mathematical thinking. New problems and investigative possibilities have emerged with the use of computer algebra systems and dynamic geometry software in pedagogic scenarios. Fallibilistic trends in philosophy of mathematics have been consolidated with focus on heuristics and challenged more strict notions of "formal proof" or "mathematical true" (e.g., the four colors theorem).
- Online distance education has offered new possibilities for in-service and pre-service mathematics teacher education (Gadanidis and Borba 2008)
- The internet has become an actor in mathematics classrooms and reorganized (a) the nature and structure of the mathematical content explored in schools or the design of lesson plans and (b) the nature of the collaboration and power relations between teachers and learners.
- The internet has a potential to make mathematics popular or accessible as a social endeavor. School mathematics usually stays inside the classrooms. Students do not have conversations with their relatives and friends about their favorite math ideas as they do when they talk about their favourite song or TV show. When parents ask to their children "what did you learn in math today?", typical responses are "nothing" or "I don't know" (Gadanidis 2009). As the Internet has a potential for democracy, the Internet has the virtual conduction to become a global stage in which students and teachers. The use of arts and the production of digital texts are fundamental to consolidate such a pedagogic/social practice.
- The genre of the online communication is similar to the genre of the performance (art). It involves multiple modes of communication, improvisation, and interaction with the "audience".

We conceptualize the *cyberspace* is a privileged educational nexus for creativity and collective intelligence. Levy (2001) defines cyberspace as the space of communication opened by the world interconnection of computers and memories of computers. This space is unique, because digital codifications shape the plastic, interactive, hypertextual, multimodal, and virtual nature of information in this context. Levy (2001) also defines *cyberculture* as the set of (materials and intellectuals) technologies, practices, attitudes, modes of thinking, and values developed through the growth of the cyberspace. The cyberculture redefines the notions of economy and knowledge, bringing up new possibilities to several areas such as education and the arts. Levy (2001) claims "the genres of cyberculture are similar to performance art, such as dance or theatre [or] the collective improvisations of jazz, the commedia dell'arte, or the traditional poetry competitions of Japan" (p. 135). Interestingly, Levy (2001) uses the term *cyberart* to discuss the artistic-aesthetic dimension of cybercultures, suggesting the possibilities for (collective) collaboration and the continuous creation as a fundamental aspect of cyberart. In other words,

> The virtual work is 'open' by design. Every actualization reveals a new aspect of the work ... Thus the creation is no longer limited to the moment of the conception or realization; the virtual system provides a machine of generating events. (Levy 2001, p. 116)

In the *Math + Science Performance Festival* (www.mathfest.ca), students, teachers and artists have shared videos in which they use the performance arts (e.g. music, drama, and poetry) to communicate their mathematical ideas. These videos are conceptualized as *digital mathematical performances* (Gadanidis and Borba 2008; Scucuglia 2012). Gadanidis and Geiger (2010) have referred to the Festival as "one example that helps bring the mathematical ideas of students into public forums where it can be shared and critiqued and which then provides opportunity for the continued development of knowledge and understanding within a supportive community of learners" (p. 102). Gadanidis and Geiger (2010) also posit the Festival "offers a glimpse into how collaboration in mathematics learning might be extended to include math performance, or perhaps how collaboration in a media-rich digital environment might be reconceptualized as collaborative performance" (p. 101).

In fact, from a narrative point of view (Bruner 1996), when students produce texts (such a video file) of a skit or a song performed in the classroom to produce a digital mathematical performance for the Festival, they are not only presenting mathematical ideas to their classmates and teachers. They are performing, communicating and representing their mathematical activity, learning, and discourses for a wide audience, because, potentially, the digital performances will be publicly available on the Festival's website. Both the classroom and the cyberspace are social/cultural settings in which students, teachers, and other agents interact, collaborate, and produce meaning and knowledge. The playful nature of digital mathematical performances offers ways of expressing ideas collaboratively, with creativity and imagination. The playfulness may also help students to make sense of mathematics through narrative because when they produce a digital mathematical performance they are seeking to communicate a mathematical story through a digital narrative/text to the audience. The process of producing a digital mathematical narrative to be published is a process in which (elementary school) students construct identities as *performance mathematicians* (Gadanidis et al. 2008; Scucuglia 2012).

Mauricio Rosa, an associate member of GPIMEM conducted part of his doctoral research under the supervision of Steve Lerman, when a visiting PhD student at the London South Bank University. In his doctoral thesis developed in our research group GPIMEM, Rosa (2008) explores the relations between the construction of online identities and the teaching and learning of calculus in an online course, when pre-service teachers perform role play games (RPG). Rosa and Lerman (2011) re-examine these data focusing on issues about research methodology. According to the authors:

> (a) [the] cyberspace is a natural environment in an online RPG context; (b) the playful process in online learning in mathematics education brings important new aspects to our understanding of mathematical knowledge as a social construction; (c) the investigation becomes a game; (d) research subjects are who they want to be while they are in flow, that is, there is intentionality; (e) the challenge of research methodology inside cyberspace must be faced by researchers; and (f) the researcher needs to consider those different identities as integral to the research process. (Rosa and Lerman 2011, p. 69)

Humans-with-Digital-Technology: Multimodality

Communication is a fundamental endeavour within sociocultural perspectives that supports mathematical classroom activity. Lerman (1998, p. 40) states that:

> Learners come to the classroom as persons of multiple, overlapping subjectivities.

Different aspects of those subjectivities are called up by different aspects of the practices of the classroom, and are expressed through identities of powerfulness or powerlessness.

> At the same time, new subjectivities are constituted in the social relationships and forms of communication which make up the activities of the classroom. Rather than the intension of teaching mathematics as the handing over, or the individual construction, of ultimately decontextualized mathematical concepts by the teacher or by the pupil respectively, teaching can be conceived of as enabling pupils to become mathematical actors in the classroom and beyond. The goals and needs of pupils, and the ways of behaving and speaking as mathematicians, become the focuses of the teacher's intentions. (Lerman 1998, p. 40)

Issues on mathematical communication also involve the socio-political context of mathematics classroom (Lerman and Zevenbergen 2004). We do recognize that students bring very different discursive rules and practices into schools and such a process "influence how they act and how actions are interpreted" (Lerman and Zevenbergen 2004, p. 32). That is,

> In considering the different discursive backgrounds of students, teachers' perceptions of their students' learning styles – that frequently correlate with the social background of the students -, and the ways in which classrooms and curricula are organized for students depending on their backgrounds, it is also important to take into account interactions within a classroom. (Lerman and Zevenbergen 2004, p. 33)

We have argued that the nature of communication based on the use of Internet is multimodal. In some scenarios of our research, students have produced multimodal texts in their classrooms to disseminate their mathematical ideas in the cyberspace. Scucuglia (2012), for instance, used the notion of multimodality in literacy to form a lens to interpret how students-with-media produce digital mathematical performances, that is, to analyze the role of digital technology in shaping students'

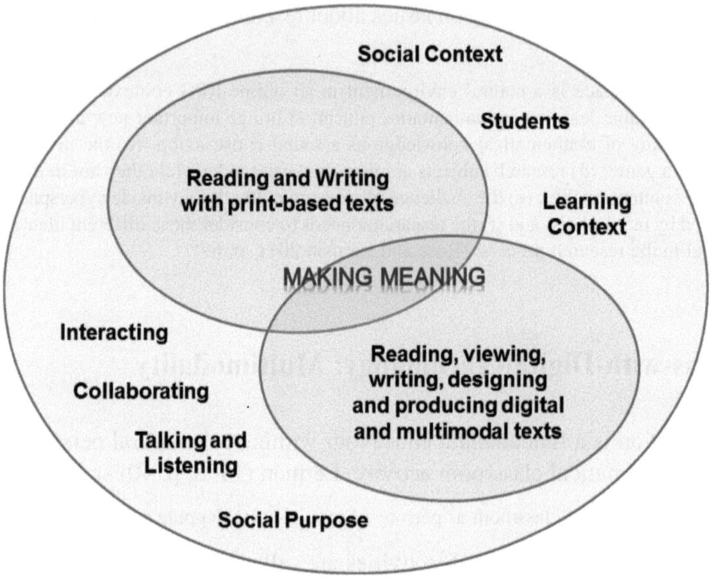

Fig. 15.2 Classroom interaction in a multimodal perspective (Walsh 2011)

mathematical thinking when they produce digital narratives using the performance arts to communicate their mathematical ideas.

In our research (Gadanidis et al. in press), we have used a model proposed by Walsh (2011) (see Fig. 15.2) to emphasize the mathematics classrooms as *social* environments with potential to form rich scenarios for multimodal learning environments when students-and-teachers-with-digital-technology produce digital mathematical performances. Walsh (2011) theorizes how classrooms can become multimodal learning environments when students interact, collaborate, and produce multimodal texts in schools. Walsh actually emphasizes the role of inter-textuality (the combination of print-based and digital/multimodal texts) and dialogue in meaning production within educational and social purposes.

Let us clarify what we mean by multimodality. Pahl and Rowsell (2005) posit that the word *multimodal* "describes the way we communicate using a number of different modes to make meaning" (p. 27). Rowsell and Walsh (2011) state that "multimodality is the field that takes account of how individuals make meaning with different kinds of modes" (pp. 55–56). According to Walsh (2011), *multimodality* is "a study of the communicative process, particularly how meaning is communicated through different semiotic or meaning-making resources and in different social contexts" (p. 105).

> Multimodality as in comprehension and competence with language through a variety of modes such as image, sound, touch, multi-dimensions, is the principle upon which digital environments work. This principle of multimodality needs to be understood for educators to apply and assess new modes of learning as a part of everyday classroom practice. (Rowsell and Walsh 2011, p. 54)

According to Gadanidis et al. (2011), "the use of multimodal expression changes the feel of the learning environment" (p. 425). In online environments, students and teachers can use text, drawings, and images, and various tools and representations.

> Different modalities – aural, visual, gestural, spatial, and linguistic – come together in one surround in ways that reshape the relationship between printed word and image or printed word and sound. Thinking with and communicating through multiple representations is a common expectation in current mathematics curriculum reform documents. (Gadanidis et al. 2011, p. 425)

Kress (2003) posits that "*m*ode is the name for a culturally and socially fashioned resource for representation and communication" (p. 45). That is, modes are "the various forms used to construct signs" (Kress 1997, p. 7). Pahl and Rowsell (2005) state that "a mode could be visual, linguistic, aural, or tacit" (p. 27). Authors like Jewitt (2006) argue that the *modalities* are aural, visual, gestural, spatial, and linguistic. The New London Group (1996) discusses language within multiliteracies based on the notion of *design*, that is, "a language for talking about language, images, texts, and meaning-making interactions ... [including] the key terms 'genres' and 'discourses,' and a number of related concepts such as voices, styles, and probably others" (p. 77).

Based on these notions, in his study, Scucuglia (2012) analyzed digital mathematical performance produced by elementary school students from Canada. The performances are available at www.mathfest.ca. Scucuglia defines a digital mathematical performance as a multimodal text/narrative (e.g., a video or a virtual learning object) in which one uses the performance arts to communicate their mathematical ideas. As part of the findings of his study, Scucuglia states that:

> The multimodal nature of [students' digital mathematical performances] is one of its most significant pedagogic attributes. Mathematics is traditionally communicated through print-based texts through the use of writing, charts, diagrams, and graphs. Digital media affordances offer ways to represent mathematical ideas through multiple modes, which adds non-usual layers of signs in communicating mathematics (e.g. audio, gestures, space). However, multimodality does not guarantee the conceptual nature of the idea explored in the [performances]. (Scucuglia 2012, p. 216)

Moreover, when students produce digital narratives, they are immersed in contexts in which they can see mathematics as stories (Gadanidis and Hoogland 2003). The synthesis between these two different modes of thinking – the paradigmatic and the narrative – to use Bruner's terms, offers ways to students to address emotions and sensations to their mathematics discourses. In doing so, students incorporate their social and cultural backgrounds into these discourses and develop communication skills (Scucuglia et al. 2011, Scucuglia 2012). We do not think that all mathematics should be communicated through digital performance, but the production of multimodal mathematical texts with emphasis on the arts is a possibility to bring representational diversity and aesthetics into the pedagogic practice of mathematics. It offers ways to challenge instructional discourses that seek "to control the content of the mathematics lesson" (Lerman and Zevenbergen 2004, p. 35).

Conclusions

We present our final remarks in terms of theory and practice. We believe that the pragmatic dimension of the use of ICT in mathematics education is not as central in Steve Lerman's work as is the theorization of sociocultural perspectives for mathematical activity. Steve uses ICT to show how theories may work in practice. In contrast, the research developed by GPIMEM shows examples and possibilities of actual use of ICT for mathematics teaching and learning and uses its research with students and teacher in order to contribute to social cultural perspectives with constructs such as humans-with-media, inter-shaping relationship and digital mathematical performance. However, in both cases, we do see theory and practice in reciprocal synergy, that is, theoretical lenses being refined based on teaching and learning experiences and pedagogic practices being conducted and reorganized based on the theoretical refinements. We believe that both research – the one developed by Steve Lerman and colleagues and the one developed by GPIMEM – has in common the notion that historically dated technology, such as Internet, may change the way communication works. Since both research approaches believe that communication is fundamental for meaning make, it can inferred that digital technology – using our terminology – is an active actor in meaning make and an actor in the process of making knowledge historically dated.

References

Borba, M. C. (1993). *Students' understanding of transformations of functions using multirepresentational software*. Doctoral dissertation, Cornell University, USA. Lisbon: Associação de Professores de Matemática.

Borba, M. C. (2004). Dimensões da educação matemática à distância [Dimensions of distance mathematics education]. In M. A. V. Bicudo & M. C. Borba (Eds.), *Educação Matemática: Pesquisa em movimento* [Mathematics education: Research in movement]. São Paulo: Editora Cortez.

Borba, M. C. (2009). Potential scenarios for Internet use in the mathematics classroom. *ZDM, 41* (4), 453–465.

Borba, M. C., Malheiros, A. P. S., & Zulatto, R. B. A. (2010). *Online distance education*. New York: Sense Publishers.

Borba, M. C. (2012). Humans-with-media and continuing education for mathematics teachers in online environments. *ZDM, 44*(6), 801–814.

Borba, M. C., & Pentado, M. G. (2001). *Informática e Educação Matemática* [Informatics and mathematics education]. Belo Horizonte: Editora Autêntica.

Borba, M. C., & Villarreal, M. E. (2005). *Humans-with-media and the reorganization of mathematical thinking*. New York: Springer.

Borba, M., & Zulatto, R. (2010). Dialogical education and learning mathematics online from teachers. In R. Leikin & R. Zazkis (Eds.), *Learning through teaching mathematics*. New York: Springer.

Borba, M. C., Malheiros, A. P. S., & Zulatto, R. B. A. (2010). *Online distance education*. Rotterdam: Sense Publishers.

Bruner, J. S. (1996). *The culture of education*. Cambridge, MA: Harvard University Press.

Cobb, P. (1994). Where is the mind? Constructivist and sociocultural perspectives on mathematical development. *Educational Researcher, 23*(7), 13–20.

Crisan, C., Lerman, S., & Winbourne, P. (2006). Mathematics and ICT: A framework for conceptualising secondary school mathematics teachers' classroom practices. *Technology, Pedagogy and Education, 16*(1), 21–39.
Crisan, C., Lerman, S., & Winbourne, P. (2007). Mathematics and ICT: A framework for conceptualising secondary school teachers' classroom practices. *Technology, Pedagogy and Education, 16*(2), 21–39.
Doll, W. (1993). Curriculum possibilities in a "post-future". *Journal of Curriculum and Supervision, 8*(4), 277–292.
Gadanidis, G. (2009). I heard this great math story the other day! *Education Canada, 49*(5), 44–46.
Gadanidis, G., & Borba, M. C. (2008). Our lives as performance mathematicians. *For the Learning of Mathematics, 28*(1), 44–51.
Gadanidis, G., & Geiger, V. (2010). A social perspective on technology enhanced mathematical learning – From collaboration to performance. *ZDM – The International Journal on Mathematics Education, 42*(2), 91–104.
Gadanidis, G., & Hoogland, C. (2003). The aesthetic in mathematics as story. *Canadian Journal of Science, Mathematics, and Technology Education, 3*(4), 487–498.
Gadanidis, G., Hughes, J., & Borba, M. C. (2008). Students as performance mathematicians. *Mathematics Teaching in the Middle School, 14*(3), 168–176.
Gadanidis, G., Hughes, J., & Cordy, M. (2011). Mathematics for gifted students in an arts- and technology-rich setting. *Journal for the Education of the Gifted, 34*(3), 397–433.
Gadanidis, G., Borba, M. C., & Scucuglia, R. (in press). Language diversity and new media: Issues of multimodality and performance. In P. Clarkson & M. C. Domite (Eds.), *ICMI study 21: Mathematics education and language diversity*. Springer.
Goos, M., Galbraith, P., Renshaw, P., & Geiger, V. (2000). Reshaping teacher and student roles in technology-enriched classrooms. *Mathematics Education Research Journal, 12*(3), 303–320.
Jewitt, C. (2006). *Technology, literacy and learning: A multimodal approach*. New York: Routledge.
Kress, G. (1997). *Before writing: Rethinking the paths to literacy*. New York: Routledge.
Kress, G. (2003). *Literacy in the new media age*. London: Routledge.
Laborde, C. (2000). Dynamic geometry environments as a source of rich learning contexts for the complex activity of proving. *Educational Studies in Mathematics, 44*(1), 151–156.
Lerman, S. (1996). Intersubjectivity in mathematics learning: A challenge to the radical constructivist paradigm? *Journal for Research in Mathematics Education, 27*(2), 133–150.
Lerman, S. (1998). *The intension/intention of teaching mathematics*. Keynote to MERGA conference. http://www.merga.net.au/documents/Keynote_Lerman_1998.pdf
Lerman, S. (2004). Researching numeracy teaching and learning with ICT: Facing the problems of innovation. In I. J. Putt, R. Faragher, & M. McLean (Eds.), *Mathematics education for the third millennium: Towards 2010 (Proceedings of the 27th annual conference of the Mathematics Education Research Group of Australasia* (pp. 618–621). Sydney: MERGA.
Lerman, S., & Zevenbergen, R. (2004). The socio-political context of the mathematics classroom: Using Bernstein's theoretical framework to understand classroom communications. In P. Valero & R. Zevenbergen (Eds.), *Researching the social dimensions of mathematics education: Theoretical, methodological and practical issues*. Dordrecht: Kluwer.
Lerman, S., & Zevenbergen, R. (2006). Maths, ICT and pedagogy: An examination of equitable practice across diverse settings. In J. Novotna, H. Morova, M. Kratka, & N. Stehlikova (Eds.), *Mathematics in the centre: Proceedings of the 30th conference of the international group for the Psychology of Mathematics Education* (Vol. 4, pp. 49–56). Prague: PME.
Lerman, S., & Zevenbergen, R. (2007). Interactive whiteboards as mediating tools for teaching mathematics: Rhetoric or reality? In *Proceedings of the 31st conference of the international group for the Psychology of Mathematics Education*. Seoul: PME.
Levy, P. (1993). *Tecnologias da Inteligência: O futuro do pensamento na era da informática* [Technologies of intelligence: The future of thinking in the informatics era]. Rio de Janeiro: Editora 34.
Levy, P. (1997). *Collective intelligence: Mankind's emerging world in cyberspace*. New York: Basic Books.
Levy, P. (1998). *Becoming virtual: Reality in the digital age*. New York: Plenum Press.

Levy, P. (2000). *Cibercultura* [Cyberculture]. Rio de Janeiro: Editora 34.
Lévy, P. (2001). *Cyberculture* (Electronic mediations). Minneapolis: University of Minnesota Press.
Mazzi, L. C., Siqueira, M. N., & Borba, M. C. (2012). As possibilidades e limitações do uso da lousa digital na Educação Matemática [Possibilities and limitations on the use of digital board in mathematics education]. *Perspectivas da Educação Matemática* [Perspectives of Mathematics Education], *5*(1), 07–18.
Noss, R., & Hoyles, C. (1996). *Windows on mathematical meaning. Learning culture and computers*. Dordrecht: Kluwer.
Pahl, K., & Rowsell, J. (2005). *Literacy and education: Understanding the new literacy studies in the classroom*. Thousand Oaks: Sage.
Papert, S. (1980). *Mindstorms: Children, computers and powerful ideas*. New York: Basic Books.
Rosa, M. (2008). *A Construção de identidades online por meio do Role Playing Game: relações com o ensino e aprendizagem de matemática em um curso à distância* [Online identity construction through role playing game: Relations with teaching and learning mathematics in a distance course]. Doctoral dissertation, Sao Paulo State University, Rio Claro, SP, Brazil.
Rosa, M., & Lerman, S. (2011). Researching online mathematics education: Opening a space for virtual learner identities. *Educational Studies in Mathematics, 78*(1), 69–90.
Rowsell, J., & Walsh, M. (2011). Rethinking literacy education in new times: Multimodality, multiliteracies, & new literacies. *Brock Education, 21*(1), 53–62.
Scucuglia, R. (2012). *On the nature of students' digital mathematical performances: When elementary school students produce mathematical multimodal artistic narratives*. Germany: Verlag/LAP Lambert Academic Publishing.
Scucuglia, R., Gadanidis, G., & Borba, M. C. (2011). Thinking collectives and digital mathematical performance. In M. Setati, T. Nkambule, & L. Goosen (Eds.), *Proceedings of the ICMI study 21 mathematics and language diversity* (pp. 348–355). Sao Paulo, Brazil.
Tall, D. (1991). *Advanced mathematical thinking*. Dordrecht: Kluwer Academic Publishers.
The New London Group. (1996). A pedagogy of multiliteracies: Designing social futures. *Harvard Educational Review, 66*(1), 60–92.
Tikhomirov, O. K. (1981). The psychological consequences of computerization. In J. Wertsch (Ed.), *The concept of activity in Soviet psychology*. Armonk: M. E. Sharpe.
Vygotsky, L. S. (1978). *Mind and society: The development of higher psychological processes*. Cambridge, MA: Harvard University Press.
Walsh, M. (2011). *Multimodal literacy: Researching classroom practice*. Sydney: e:lit, Primary Teachers Association of Australia.
Zevenbergen, R., & Lerman, S. (2005). Classroom interactions: Scaffolding students learning with ICTS. In M. Goos, C. Kanes, & R. Brown (Eds.), *Proceedings of the 4th international conference of the Mathematics Education and Society Group*. Brisbane: Centre for Learning Research.
Zevenbergen, R., & Lerman, S. (2006). Maths, ICT and pedagogy: An examination of equitable practice in diverse contexts. In J. Novotna, H. Moraova, M. Kratka, & N. Stehlikova (Eds.), *Proceedings of the thirtieth meeting of the International Group for the Psychology of Mathematics Education* (Vol. 4, pp. 49–56). Prague: Charles University.
Zevenbergen, R., & Lerman, S. (2007). Interactive Whiteboards as mediating tools for teaching mathematics: Rhetoric or reality? In J.-H. Woo, H.-C. Lew, K.-S. Park, & D.-Y. Seo (Eds.), *Proceedings of the thirty-first meeting of the International Group for the Psychology of Mathematics Education* (Vol. 3, pp. 169–176). Seoul: The Korea Society of Educational Studies in Mathematics.
Zevenbergen, R., & Lerman, S. (2008). Learning environments using interactive whiteboards: New learning spaces or reproduction of old technologies? *Mathematics Education Research Journal, 20*(1), 107–125.

Chapter 16
Troubling Mathematics "Learners"

Brent Davis

Introduction

Many of Stephen Lerman's writings have dealt with the simultaneous importance of considering diverse theoretical offerings while avoiding "incautious complementarities." In this chapter I explore Steve's manners of framing these discussions, focusing on key orienting metaphors and issues. In the process, I offer (perhaps incautious) complementary readings of theoretical possibilities within mathematics education.

There are many qualities about Stephen Lerman's work that I admire, among them his fastidious attention to conceptual detail and his broad familiarity with the theoretical terrain. But easily the most impressive aspect for me is his capacity to agree while disagreeing – to foreground the insight at the same time as he reveals the blind spot.

This ability is perhaps best demonstrated across his nuanced explorations of theories of learning and their relevance for mathematics learning and teaching. To my reading, the ability seems to be associated with an attitude that "no one is completely wrong," which in turn is hinged to a talent for discerning exactly what people are talking about when they invoke such taken-as-shared notions as "mathematics," "truth," and "learning."

As he demonstrated in his 2002 ESM article (Lerman 2002), the metaphor of *zooming in* and *zooming out* is particularly useful for making these sorts of discernments. Tethered to Rogoff's (1998) identification of three foci of analysis in sociological research – the personal, the interpersonal, and the cultural/contextual – Lerman uses the metaphor to embrace the simultaneity of different levels of activity, without collapsing them into or reducing them to one another. By zooming in and zooming out, one can address the particular integrities of different levels of

B. Davis (✉)
Faculty of Education, University of Calgary, Calgary, AB, Canada
e-mail: brent.davis@ucalgary.ca

P. Gates, R. Jorgensen (Zevenbergen) (eds.), *Shifts in the Field of Mathematics Education*, DOI 10.1007/978-981-287-179-4_16

phenomena and thus avoid conflating such theories as the individual-focused radical constructivism and more collective-minded social constructivisms. As he noted, "incompatibilities lurk in incautious complementarities" (p. 88). This warning is a common one in Steve's writings, elsewhere expressed as a common error (e.g., "… the assumption of complementarity [between radical constructivism and social constructivism] leads to incoherence" – Lerman 1996, p. 133) and a frequent shortfall (e.g., "It is no easy matter to join together different theories and it is done unsatisfactorily far too often" – Lerman 2006, p. 9).

It is an important caution, and one that Steve has voiced to me personally more than once in not-so-subtle criticisms of my own tendency to seek out complementarities among theoretical offerings. With that hazard warning acknowledged, however, in this chapter I once again set out to do precisely what Steve advises one not do: I attempt to reconcile some of the more prominent theories of learning as represented in current mathematics education literature.

Seeking an Appropriate Metaphor

In addition to connecting his usage of the zooming metaphor to Rogoff's foci or planes of analysis, Steve noted a resonance to Cobb's (2000) discussion of "grain size" and the more commonly encountered notion of "unit of analysis." However, with regard to these latter elements, Steve argued that *zooming* offers a more fecund entry to the task of bringing …

> … the macro and the micro together, one that enables us to examine how social forces such as a liberal-progressive position, affect the development of particular forms of mathematics thinking. (Lerman, op cit, p. 89)

Those points I find compelling, and have little difficulty embracing. Whereas *grain size* and *unit of analysis* seem to compel a singularity of focus (onto the grain or the unit), *zooming* invites one to be mindful of the always-multi-plied (i.e., many-layered) complexity of mathematics teaching and learning.

In Fig. 16.1 I've attempted to summarize some of the key insights and entailments of the *zoom* metaphor, as I understand them. For me, this graphic is particularly useful for highlighting the necessary integrities of the theories we use and the phenomena they describe. It is also useful for foregrounding the realization that understandings and interpretations are inevitably partial – in the double sense of partial = fragmentary and partial = biased.

As well, the act of zooming implies a conscious awareness of selection, thus offering a reminder of the complicity of the observer in the phenomenon observed. As Steve phrases it, "the particular focus creates the object of research" (p. 90).

On the matters of the utility of the *zoom* metaphor and the complicity of the researcher, I am in agreement. Embracing the advice that Steve offers based on these insights, however, is another matter. He knits them into grounds for the assertion "that neither complementary nor emergent views can achieve this

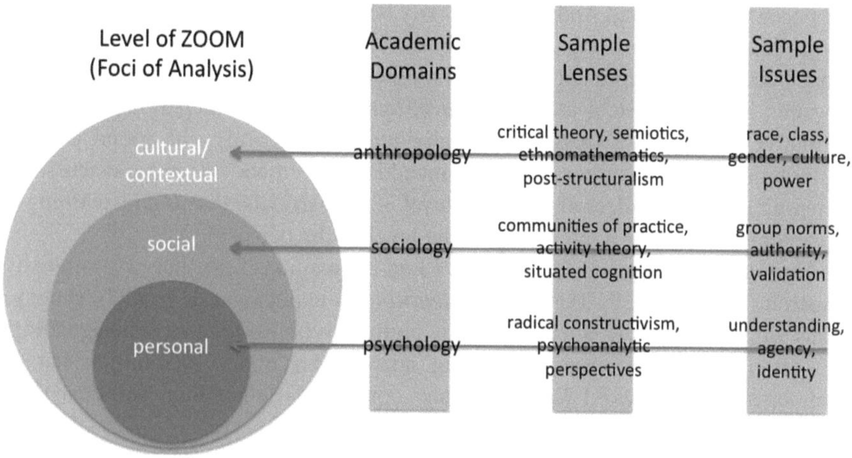

Fig. 16.1 Some entailments of the ZOOM metaphor

integration" (ibid.). I disagree. And while this disagreement is no doubt tethered to discordant understandings of some key words (including *complementary* and *emergent*), I suspect it is mostly tied to the orienting metaphor. While I agree that *zooming* represents an improvement to *grain size* and *unit of analysis*, it doesn't do much to elaborate the span of perceptible forms. That is, the discussion continues to be bookended by the personal at one extreme and the cultural at the other. It is more consciously multifaceted, but it nonetheless addresses only elements that are immediately available to human consciousness, that is to those aspects of our experience that we can be (but very often are not) aware of in our daily acting.

It is against this backdrop that I propose a different metaphor, one that extends the bookends of the discussion in both micro and macro directions. As I develop in the next section, I seek complementarities about theories of learning through a metaphor drawn from the domain of complexity science, the study of emergence. Phrased differently, precisely contrary to Steve's advice, I look to a fusion of complementary and emergent views to articulate some manner of integration.

What Is Learning? Where Does It Happen? Who/What Does It?

Throughout his significant opus, Steve regularly challenged assumptions about where and how learning happens. One question that he doesn't seem to engage, however, is what or who is doing the learning. That is, what's a learner?

A few decades ago this question would've been heard as absurd in educational circles. "Learner" was synonymous with "student," "pupil," or "individual." These words were used interchangeably to refer to the person – the *psyche* – who was supposed to be learning. And that's certainly how I interpret Steve's use of the word across his publications (e.g., Lerman 1996, 1998, 2005). In other words, Steve's work betrays a commitment to a psychological construct of learner = individual, revealed in the consistent usages of *learn*, *learning*, and *learner* to refer to transformative actions and occasions of individuals. Importantly, the social and the cultural are deeply incorporated (i.e., embedded and embodied) in his discussions of learning, but regardless of the zoom level of the particular discussion, learning is a phenomenon that he applies only to the level of the personal.

Among other usages of *learner* that I find compelling is one out of complexity research (e.g., Mitchell 2009), where *learning* is understood as roughly synonymous with *evolution*. It is used by many complexivists to refer to self-regulating, self-determining, coherence-maintaining, adaptive behavior – a definition that, in turn, opens up the word *learner* to any coherence-maintaining, self-changing, emergent system.

The meaning of the word *emergent* is very specific here. An emergent phenomenon in one in which similar but nonetheless diverse elements coalesce into a coherent, discernible unity that cannot be reduced to the sum of its constituents. The emergent unity arises in the activities of the agents it comprises, but transcends them. Brains are more than collections of neurons, humans are more than assemblages of organs, societies are more than groups of people. Each of these emergent, ecosystemic forms is a *learner*, metaphorically speaking, and *learning* is recast as any event by which a coherent system transforms in a manner that extends its viability while preserving its continuity/identity.

Importantly, a learner need not be a physical entity. There are other possibilities, including ideational learning systems. For example, it is not uncommon to encounter suggestions that mathematics itself "is a living, breathing, changing organism . . ." (Burger and Starbird 2005, p. xi) or that it "emerges as an *autopoietic* [i.e., self-creating and self-maintaining] system" (Sfard 2008, p. 129). Mathematics is a learner.

In my experience, this metaphorical extension of *learner* is heard by many as a radical suggestion. However, it is not an altogether new idea. Its history has been traced in some detail elsewhere (e.g., Davis and Sumara 2006; Davis et al. 2008), and it is important to note that it is a notion that has been creeping silently, but pervasively into the mathematics education literature. In particular, observations and analyses of collective cognition have a fairly deep history, with discussions typically framed in terms of collective cognition of a social grouping. Such analyses have been provided by Bauerserfeld (1992), Cobb (1999), Kieren (2000, 2001), Sfard and Kieran (2001), and Zack and Graves (2001), to name a few.

This work has paralleled a broader discussion of "knowledge building" in the field of education (e.g., Scardamalia and Bereiter 2003), a conceptual and pragmatic movement that is oriented toward the collaborative production of understandings that go beyond the level of the most knowledgeable individual in the learning group. Included among the core principles of knowledge building are the need to create an idea-rich environment, the convictions that all ideas are improvable, the requirement to gather and weigh evidence, and the centrality of participation.

Although rooted in the social sciences literature, the knowledge-building frame is readily aligned with an emergentist frame – and, indeed, the word emergent appears frequently in the associated literature (although not always linked to the a complexity frame).

Even though use of the term emergent is common across many analyses and discussions, researchers frequently stop short of identifying classroom collectives or knowledge domains as learning systems. While they recognize that social norms, interaction patterns, and mathematical ideas emerge and evolve, there remains a strong tendency to describe these phenomena at the level of interacting agents, and not as properties of an emergent collective unity/learner. By contrast, outside of education, there seems to be a much greater readiness to extend the notion of learner to many nested forms. For example, Jablonka and Lamb (2005) described the knowledge of an agent in terms of inheritance (i.e., learning that has previously occurred at another level of organization and, e.g., encoded in a genome, written in an authoritative test, or knitted into a neurological network) and adaptation (i.e., learning that occurs at the level of the agent as conditioned by inheritance and as occasioned by context) that work simultaneously across four scales: genetic, epigenetic, behavioral, and symbolic.

These four scales (or levels of learning) are clearly related to, but different from Rogoff's three foci of analysis (i.e., personal, social, cultural/contextual). Both lend themselves to the *zoom* metaphor, but to really appreciate their difference it's important to recognize that Jablonka and Lamb (and other complexivists) do not speak in terms of shifting foci or moving between planes of analysis, but of actual nested entities or co-implicated learners. In the process, they contribute to the possibility of a much-elaborated conversation that includes both the sub-personal and the more-than-human, as I attempt to illustrate in Fig. 16.2.

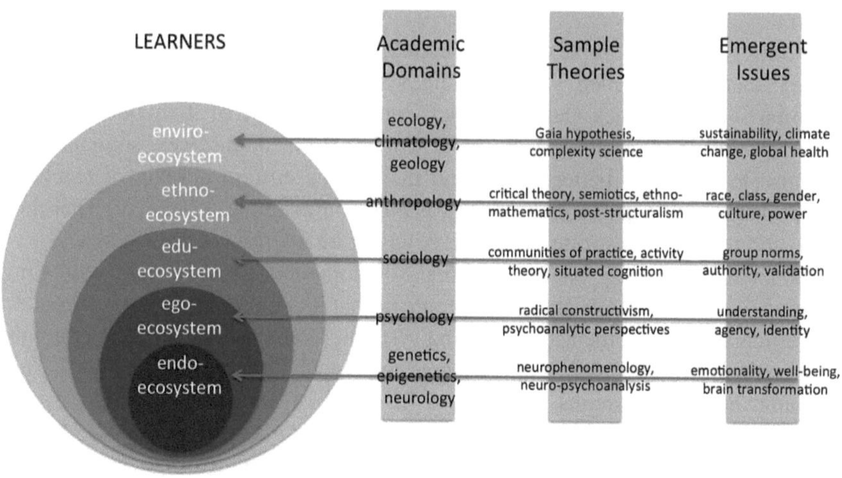

Fig. 16.2 Some entailments of the LEARNERS metaphor

These elaborations into the not-directly-perceptible spaces of the sub-personal and more-than-human are recent but vital additions to discussions of mathematics learning. They are enabled – and perhaps necessitated – by two major developments. Firstly, advances in technology and accumulations of data have made it possible to re-scale the timeframes of phenomena on these levels in ways that make their learnings perceptible to us. Secondly, particularly on the more-than-human level, some learnings/transformations, such as climate change and ocean acidification, have accelerated to a pace that they are directly perceptible. In the process, a new host of critical issues have arisen for educators, including global health and sustainability.

A Note on Grammar and Logic

In a 2006 article, Steve drew on Bernstein (1999) as he commented on the contrast between the "strong grammar" of mathematics and the sciences and the "weak grammar" of education and other humanities (Lerman 2006). A field with a weak grammar has "a conceptual syntax not capable of generating unambiguous empirical descriptions" (p. 9), in contrast to one with a strong grammar.

Steve saw this as both boon and bane for mathematics educators. On the positive side, "the weakness of the grammars in mathematics education research is more likely to enable communication and even theory-building across different discourses" (ibid.). On the negative side, a weak grammar is necessarily a major obstacle in the development of accepted knowledge. Even neuroscience, arguably with the strongest grammar of any learning-oriented field, is saddled by a weak grammar, as demonstrated in its need to qualify every "finding" with a commentary on the radical variations that exist among brains (see, e.g., Davidson and Begley 2012).

Working from a complexity perspective, I would argue that any domain of inquiry that is concerned with learning systems will inevitably lack a strong grammar – and that lack can be, in fact, construed as one of its major strengths. Learning systems are slippery forms. They transform themselves, sometimes in response to efforts to better understand them. (The current domain of teachers' disciplinary knowledge of mathematics is a cogent example, I believe.) That they learn, and thus belie a strong grammar, need not be interpreted as a weakness.

Rather than focusing on the grammars of domains of inquiry, and following Maturana and Varela (1987), I would contend that it is more important to attend to the particular logics of the actual learning systems. Domains with a strong grammar are associated with operations that are governed by highly prescriptive logics. That is, truths in these domains are defined in terms of what must be done – the conditions that must be met, the contexts in which an assertion is true, the limitations of claims, and so on. By contrast, domains that work with a weak grammar are enabled by a proscriptive logic, in which constraints are articulated not in terms of what must be done, but in terms of what is not allowed. In brief, only those actions that threaten the viability of the learning form are prohibited.

One can readily observe the fecundity of a proscriptive logic in games (where rules are principally expressed in terms of what you're not allowed to do, such as moving a bishop laterally or double-hitting a volleyball). It is even present in domains with a strong grammar. The many branches of mathematics, for example, do not spring from its prescriptive demands but its proscriptive possibilities. As long as it's not prohibited, it's a possible space of action.

I use this point to segue to a brief discussion of my real interest: educational pragmatics. Personally, I find the zooming metaphor and the pervasive anxiety over the necessarily weak grammar of education to be limiting constraints in discussions of educational possibility. Certainly they afford important insights, but they seem to compel qualifications and apologies as they pull attentions toward the inevitable exception or the eclipsed detail. By contrast, the combination of a *learner* metaphor and an awareness of the expansive dynamics that are afforded within systems governed by proscriptive logics presses attentions toward intricacies and possibilities. For me, these gesture more toward educational discourses – that is, discourses that seek and create learning potentials.

Toward Resolution

It's one thing to grapple with the complementarities and incompatibilities of theories. It's quite another to engage with them in a manner that speaks to the work of educators. Steve has always done both, and his opus is a powerful demonstration of how attentiveness to conceptual detail can be driven by and oriented toward a concern for pragmatic possibility. As I read his writing, the motivator always seems to be the possibilities that might be afforded for the teaching and learning of mathematics.

I share that motivation, and it is why I am so willing to flout cautions about the inadequacies of both "complementary nor emergent views." Like Steve, I see the real world of the mathematics classroom as, simultaneously, the site of contestation across theories and the necessary space of their resolution.

The word *resolution* is an interesting one, as it carries nearly opposite definitions in different fields. In Law, a resolution is a consolidating and binding court decision; in Medicine, a resolution is a subsiding or a termination of an abnormal condition; in Chemistry, a resolution is a dissociation of something into its constituent parts. That is, the word can mean either something toward "holding firmly," something toward "diminishing," or something toward "breaking apart."

On that count, I vividly recall, in the early 1990s, a fellow graduate student's delight at the use of the word *solution* to refer to mathematical problems. From Eastern Europe, this usage of the term was new to her, having encountered it before only in chemistry. For her to that point, a solution was a mixture of things; in essence, a set of potentials, depending on what was present, the surrounding conditions, and so on. She thus thought it a brilliant notion to apply to mathematical problems – seeking a solution (in the chemistry sense) seemed so much richer than looking for an answer offering a response. Needless to say, she was more than a little disappointed to learn the popular usage carried few or none of these subtle connotations.

Nevertheless, the event had a significant impact on me, and it has figured into my teaching and research ever since. It bespeaks a complexivist sensibility as it summons both the original meanings of solution and resolution (having to do with processes of reducing things into simpler forms) and their emergent definitions of "coming to a singleness of interpretation." Within the science of emergence, these aren't incompatible processes. In fact, quite the opposite, they're necessary complementarities.

So, on to a pragmatic question: What might complexity thinking say about mathematics teaching?

I'll begin by situating this discussion alongside a few terms that have popped up in the mathematics education literature over the past few decades. These words, I think, offer a window onto teaching that might be aligned with a complexivist, emergentist sensibility: *improvising*, *occasioning*, *caring*, *conversing*, *listening*, *mindful participation*, and *engaging*. These sorts of notions evoke what Donald (2002) called a "coupling of consciousnesses" – a uniquely human capacity to coordinate attentional systems and to synchronize brain function, in effect presenting the possibility of grander cognitive unities. Compared to individual capacities, collectively humans are able to keep vastly more ideas in mind, to make much more impressive connective links, and so on. As Donald develops, this possibility of collective mind (vs. collection of minds) is biologically rooted, but greatly enabled by language and other social conventions. It is, simultaneously, a neurological, personal, social, and cultural phenomenon.

Given this capacity for complex, communal cognition (i.e., for learning at multiple levels simultaneously), what is the role of the mathematics teacher? I believe a powerful possibility is offered through the metaphor of "teacher as the consciousness of the (classroom) collective."

Four critical points are necessary to make sense of this suggestion. Firstly, human consciousness depends on social collectivity, at the same time as it is personal and individual. As Donald points out, human cognitive systems (or minds) are hybrid, dependent on both an individual brain and various levels of collectivity. Secondly, applying the notion of communal cognition, the potential must exist for a classroom community to emerge into a coherent unity. It must be possible, even if it's not particularly common. Thirdly, as Donald describes, the past century of consciousness studies have demonstrated that consciousness does not control, it commentates. It doesn't direct, it orients as it engages in a continuous mash-up of history and new experience. What one knows and who one is, then, are not determined by consciousness, but they are reflected by what goes on in consciousness. Extending this detail (metaphorically) to education, just as identity and action depend on but are not determined by consciousness, communal learning in the classroom depend on but are not determined by the teacher.

The metaphor of "teacher as the consciousness of the collective," then, is a suggestion that the main work of the teaching is around prompting differential attention, selecting from among the options for action and interpretation that arise in the collective. This formulation, of course, only makes sense insofar as there are options for action and interpretation that afford selection, which bring me to the fourth critical point. I'll frame this one with a quote from Damasio (2005), a neurologist:

> [I]n the nervous system, as much as the immune system, selection from among diverse elements is more important than instruction to shaping functional structure. (p. 72)

Replace "nervous system" with "classroom" and you have an important truth about collective learning and the role of the teacher. The teacher's task in not just to select from among those possibilities that present themselves to her or his awareness. A vital preliminary task has to do with ensuring that diverse interpretive possibilities are present in the classroom.

So framed, teaching cannot be about zeroing in on predetermined conclusions. It can't be about the replication and perpetuation of the existing possible. Rather, teaching seems to be more about expanding the space of the possible and creating conditions for the emergence of the as-yet unimagined. So emergentist teaching is not about prompting a convergence onto pre-existent truths, but about the divergence into new interpretive possibilities as conditioned by inherited knowledge. The emphasis is not only on what is, but also what might be brought forth. Mathematics teaching comes to be a participation in a recursively elaborative process of opening up new spaces of possibility by exploring current spaces.

And So . . .?

I set as my task in this writing both to heed and to challenge Steve's assertion "that neither complementary nor emergent views can achieve an integration" of diverse theoretical perspectives. He may be correct, but even if that is the case, I would still argue that an attitude that *combines* complementarity and emergentism – of discrete parts contributing to grander wholes – might enable some manner of integrated insight. Complementary and emergent views needn't be positioned as poles or incompatibilities.

The above metaphorical play around the nature and place of teaching is offered as one illustrative instance of this possibility, and I selected it as an exemplar because of the obviousness of a phenomenon within a phenomenon. Teacher as consciousness of the collective is a part-and-whole construct. And it is easy to nest this construct within other phenomena, not so much by zooming but by considering other co-implicated *learners*, such as the educational system, culture, and the transcultural phenomenon of mathematics itself.

This point is a vital one. With the strong undercurrent of individuality in western sensibilities, the suggestion that attention be diverted from the individual to the group is often heard as an attempt to diminish the uniqueness of the person, whether that person is a student or the teacher. Exactly the opposite is intended here. In fact, it is precisely the uniqueness of individual contributions that define the extent of collective possibility. A group in which everyone contributes or focuses on the same thing is not likely to move beyond that singular focus. A group that offers and grapples with diversity of interpretation has potential to go to interesting new places. In other words, "the best way for a group to be smart is for each person to act as independently as possible" (Surowiecki 2004, p. xx) – where "independence" is in reference to personal choice, not individual isolation. Group-based intelligence is not rooted in a this-or-that logic. The possibilities for the individual learner and the collective learner can and should amplify one another.

And so in response – or, perhaps, in resolution – to Steve's very important and utterly relevant warning that "incompatibilities lurk in incautious complementarities" I would counter that mathematics education, as a field, risks irrelevance if it does not at least attempt a coherence across theoretical offerings. The intention is not to combine, collapse, distill, or prioritize. Nor is it to arrive at a singular truth. Rather, the goal is realized in the engagement. What might we learn by forcing the conversation?

Of course, possibilities will be dictated by the extent to which we listen to one another, and it is here that I feel Steve has made his greatest contribution and left his biggest mark on the field of mathematics education. As evidenced in all his writings, it is clear that he put forth the effort to deeply understand the theories he critiqued. More than once I have turned to one piece or another of Steve's writings to remind me of a nuance that separates a particular perspective from another. Those nuances, those openings for debate and conversation, are precisely the sites of more expansive, enabling theoretical possibilities. As long as we take the care to avoid incautious complementarities as we take the dare (as I believe we must) to articulate emergent, complementary theories.

References

Bauerserfeld, H. (1992). Integrating theories for mathematics education. *For the Learning of Mathematics, 12*(2), 19–28.

Bernstein, B. (1999). Vertical and horizontal discourse: An essay. *British Journal of Sociology of Education, 20*(2), 157–173.

Burger, E. B., & Starbird, M. (2005). *The heart of mathematics: An invitation to effective thinking* (2nd ed.). Emeryville: Key College Publishing.

Cobb, P. (1999). Individual and collective mathematics development: The case of statistical data analysis. *Mathematics Thinking and Learning, 1*(1), 5–43.

Cobb, P. (2000). The importance of a situated view of learning to the design of research and instruction. In J. Boaler (Ed.), *Multiple perspectives on mathematics teaching and learning*. Westport: Ablex.

Damasio, A. (2005). Think tank: Neuroscience – Top scientists pinpoint the critical development of the last 25 years and predict wonders yet to come. *Discover, 26*(5, May), 72–75.

Davidson, R. J., & Begley, S. (2012). *The emotional life of your brain: How it unique patterns affect the way you think, feel, and live – And how to change them.* New York: Hudson Street Press.

Davis, B., & Sumara, D. (2006). *Complexity and education: Inquiries into learning, teaching, and research.* Mahwah: Erlbaum.

Davis, B., Sumara, D., & Luce-Kapler, R. (2008). *Engaging minds: Changing teaching in complex times.* New York: Routledge.

Donald, M. (2002). *A mind so rare: The evolution of human consciousness.* New York: W.W. Norton.

Jablonka, E., & Lamb, M. J. (2005). *Evolution in four dimensions: Genetic, epigenetic, behavioral, and symbolic variation in the history of life.* Cambridge, MA: MIT Press.

Kieren, T. E. (2000). Dichotomies or binoculars: Reflections on the papers by Steffe and Thompson and by Lerman. *Journal for Research in Mathematics Education, 31*, 228–233.

Kieran, C. (2001). The mathematical discourse of 13-year-old partnered problem solving and its relation to the mathematics that emerges. *Educational Studies in Mathematics, 46*, 187–228.

Lerman, S. (1996). Intersubjectivity in mathematics learning: A challenge to the radical constructivist paradigm? *Journal for Research in Mathematics Education, 27*, 133–150.

Lerman, S. (1998). Cultural perspectives on mathematics and mathematics teaching and learning. In F. Seeger, J. Voigt, & U. Waschescio (Eds.), *The culture of the mathematics classroom* (pp. 290–307). Cambridge: Cambridge University Press.

Lerman, S. (2002). Cultural, discursive psychology: A sociocultural approach to studying teaching and learning of mathematics. *Educational Studies in Mathematics, 46*, 87–113.

Lerman, S. (2005). Learning mathematics as developing identity in the classroom. In P. Liljedahl (Ed.), *Proceedings of the annual meeting of the Canadian Mathematics Education Study Group* (pp. 3–13). Burnaby: CMESG/GCEDM.

Lerman, S. (2006). Theories of mathematics education: Is plurality a problem? *ZDM, 38*(1), 8–13.

Maturana, H. R., & Varela, F. J. (1987). *The tree of knowledge: The biological roots of human understanding.* Boston: Shambhala.

Mitchell, M. (2009). *Complexity: A guided tour.* Oxford: Oxford University Press.

Rogoff, B. (1998). Cognition as a collaborative process. In W. Damon, D. Kuhn, & R. S. Siegler (Eds.), *Handbook of child psychology* (pp. 679–744). New York: Wiley.

Scardamalia, M., & Bereiter, C. (2003). Knowledge building. In J. W. Guthrie (Ed.), *Encyclopedia of education* (2nd ed.). New York: Macmillan Reference.

Sfard, A. (2008). *Thinking as communicating: Human development, the growth of discourses, and mathematizing.* New York: Cambridge University Press.

Sfard, A., & Kieran, C. (2001). Cognition as communication: Rethinking learning-by-talking through multi-faceted analysis of students' mathematical interactions. *Mind, Culture, and Activity, 8*(1), 42–76.

Surowiecki, J. (2004). *The wisdom of crowds.* New York: Doubleday.

Zack, V., & Graves, B. (2001). Making mathematical meaning through dialogue: Once you think of it, the *z* minus three seems pretty weird. *Educational Studies in Mathematics, 46*, 229–271.

Index

A

Activity theory, 2, 4–7, 9, 21, 25, 67, 69–71, 114, 158, 178, 181, 186–193, 195, 198, 221

B

Bernstein, B., 4, 6, 9–11, 48–50, 52, 53, 55, 61, 113–115, 118, 126, 140, 141, 146, 147, 158, 236

Bourdieu, P., 4–9, 12, 49, 50, 61, 115

British Congress of Mathematics Education (BCME), 50

British Society for Research into Learning Mathematics (BSRLM), 7, 37

Bruner, J., 158, 163, 224, 227

C

Capital, 5, 7–9, 160, 165

Class/social class, 5, 11, 12, 30, 33, 46–50, 52–55, 61, 65, 85, 107, 110, 113, 123–126, 128, 129, 132, 134, 163, 164, 177, 180, 207, 222

Cognitive ecology, 218, 219

Collective intelligence, 219, 220, 223

Community of inquiry, 69, 176

Complex instruction, 115–116

Complexity, 61, 64, 67, 71, 82, 125, 126, 175, 178, 198, 232–236, 238

Constructivism/constructivist
- radical, 8, 9, 21, 33, 79, 122, 151, 154–156, 158–163, 165, 172, 173, 175, 179, 180, 182, 210, 211, 216, 232
- social, 5, 8, 154–156, 158, 161, 165, 181, 207, 211, 216, 232

Cultural psychology, 9, 11, 64, 65, 182, 216

Cyber
- art, 224
- culture, 223, 224
- space, 224, 225

D

Davydov, V.V., 151, 164, 188

E

Epistemology, 63, 153, 155, 159, 165, 209, 210, 212

Equity, 2, 12, 43–55, 112–117, 211

Ethnomathematics, 61, 82

Expansive learning, 186, 196

G

Grammar
- strong grammar, 140, 210, 236, 237
- weak grammar, 10, 11, 140, 141, 144, 147, 148, 210, 236, 237

H

Hermeneutical phenomenology, 84, 209

Humans-with-media, 215–220, 222, 228

I

Identity, 6, 22, 52, 54, 67, 69–71, 78, 85, 141, 156, 166, 195, 198, 207, 211, 234, 238

Indigenous, 39, 84, 108, 113, 115–118, 127

P. Gates, R. Jorgensen (Zevenbergen) (eds.), *Shifts in the Field of Mathematics Education*, DOI 10.1007/978-981-287-179-4

Intersubjectivity, 2, 3, 156, 157, 160–162, 171–183, 216

J

Jewish, 22, 23, 31, 45, 83

K

Kibbutz, 30, 45, 49
Knowledge
  horizontal knowledge, 140
  mathematical knowledge for teaching (MKT), 139–148
  pedagogical content knowledge (PCK), 24, 142–144
  subject matter knowledge (SMK), 24, 142–145
  vertical knowledge, 140

L

Locke, J., 152

M

Marx/Marxism/Marxist, 45–47, 49, 53
Mathematics Education and Society (MES), 10, 34, 51–54, 61, 83
Multimodal, 215, 223, 225–227

P

Philosophy, 2, 3, 20, 23, 31, 32, 47, 59, 62, 63, 152, 155, 203–212, 223
Piaget, 22, 23, 33, 64, 153, 156, 159, 162–164, 172, 216
Postmodern, 54, 67, 160
Psychology of Mathematics Education (PME), 7, 10, 18, 21, 24, 32–34, 37, 38, 45, 51–54, 61, 79, 95, 96, 98, 102, 123, 151, 153–156, 158, 160–162, 164–166

R

Radical Constructivism. *See* Constructivism

S

Secondary Mathematics Individualised Learning Experiment (SMILE), 19, 20, 46
Social Constructivism. *See* Constructivism
Social justice, 2, 12, 43–48, 51, 52, 54, 80, 207, 211
Social turn, 1, 2, 4–6, 9–12, 17–26, 48, 55, 64–66, 68, 70, 75–86, 151, 158, 160, 166, 185, 186, 210
Sociocultural, 3, 17–19, 21–25, 52, 53, 59, 61, 64–71, 80, 82, 113, 114, 152, 153, 159–163, 165, 166, 171–183, 207, 210, 211, 215–219, 225, 228
Speech act, 2, 9, 10, 75–86

U

Universally distributed intelligence, 219

V

Values, 2, 4, 5, 9, 18, 26, 36, 46–49, 54, 62, 70, 84, 85, 111, 128, 175, 178, 190, 191, 199, 204, 206–208, 211, 223
Vico, G., 152–154, 157
Vygotsky, L., 9, 21–23, 25, 33, 49, 61, 65, 113, 114, 118, 123, 140, 157, 158, 160–164, 173, 176, 178, 181, 185–187, 199, 216–218

W

Wittgenstein, L., 20, 21, 23, 155, 158, 161, 187, 191, 211

Z

Zone of proximal development (ZPD), 25, 35, 123, 131, 158, 164–166, 192, 196, 217

MIX
Papier aus verantwortungsvollen Quellen
Paper from responsible sources
FSC® C105338

If you have any concerns about our products,
you can contact us on
ProductSafety@springernature.com

In case Publisher is established outside the EU,
the EU authorized representative is:
Springer Nature Customer Service Center GmbH
Europaplatz 3, 69115 Heidelberg, Germany

Printed by Libri Plureos GmbH
in Hamburg, Germany